Ernest Ingersoll

The Book of the Ocean

Ernest Ingersoll

The Book of the Ocean

ISBN/EAN: 9783337037895

Printed in Europe, USA, Canada, Australia, Japan

Cover: Foto ©berggeist007 / pixelio.de

More available books at **www.hansebooks.com**

THE
BOOK OF THE OCEAN

BY

ERNEST INGERSOLL

AUTHOR OF "KNOCKING ROUND THE ROCKIES," "THE OYSTER INDUSTRIES OF
THE UNITED STATES," "FRIENDS WORTH KNOWING," "WILD
NEIGHBORS," "THE CREST OF THE CONTINENT," ETC.

Illustrated

NEW YORK
THE CENTURY CO.
1898

THE DE VINNE PRESS.

CONTENTS

THE BOOK OF THE OCEAN

THE BOOK OF THE OCEAN

OOKING at the land, we divide the surface of the earth into eastern and western hemispheres; but looking at the water, we make an opposite classification. Encircle the globe in your library with a rubber band, so that it cuts across South America from about Porto Alegre to Lima on one side, and through southern Siam and the northernmost of the Philippine Islands on the other, and you make hemispheres, the northern of which (with London at its center) contains almost all the land of the globe, while the southern (with New Zealand as its central point) is almost entirely water, Australia, and the narrow southern half of South America being the only lands of consequence in its whole area. Observing the map in this way, noticing that, besides nearly a complete half-world of water south of your rubber equator, much of the northern hemisphere also is afloat, you are willing to believe the assertion that there is almost three times as much of the outside of the earth hidden under the waves as appears above them. The estimate in round numbers is one hundred and fifty million square (statute) miles of ocean surface, as compared with about fifty million square miles of land on the globe.

To the people whose speculations in geography are the oldest that have come down to us, the earth seemed to be an island around which was perpetually flowing a river with no further shore visible. Beyond it, they thought, lay the abodes of the dead. This river, as the source of all other rivers and waters, was deified by the early Greeks and placed among their highest gods as Oceanus, whence our word "ocean." Accompanying, or belonging to him, there grew up, in the fertile imagination of that poetic people, a large company of gods and goddesses, while men hid their absence of real knowledge by peopling the deep with quaint monsters.

"The word for 'ocean' (*mare*) in the Latin tongue means, by derivation, a desert, and the Greeks spoke of it as 'the barren brine.'"

Over these old fables we need not linger. All the myths and guess-work that went before history represented the sea as older than the land, and told how creation began by lifting the earth above the universal waste of waters. The story in Genesis is only one of many such stories.

Scientific men believe that when our planet first went circling swiftly in its orbit it was a glowing, globular mass of fiery vapors: but as time

A QUIET SEA, AND THE SUN AT MIDNIGHT.
From a photograph.

passed, the icy chill of space slowly cooled these vapors, and chemical changes steadily modified, sorted, and solidified the materials into the beginnings of the present form and character, until at last *water* came into existence. This must have been at first in the form of a thick envelop of heated vapors, impregnated with gases, that inwrapped the globe in a darkness lit only by its own fires.

After that, when further changes had come about,—let us picture it,— what deluges of rain were poured out of and down through those murky clouds where thunders bellowed and lightnings warred! At first all the rains that fell must have been turned to steam again; but by and by the

EATING AWAY THE COAST.

steady downpour cooled the shaping globe so that all the water was not vaporized, but some stayed as a liquid where it fell, and this increased in amount more and more, until finally, between the hissing core of the half-hardened planet and the dense clouds which kept out all the sunlight, there rolled the heated waves of the first ocean —an ocean broken only by the earliest ridges, like chains of islands, marking the skeletons of the continents that were to follow — an ocean sending up ceaseless volumes of steam to form new clouds.

Yet all the while the cooling of the planet went on. Now, when any heated substance cools it contracts, and the globe as a whole is no exception to the rule; but a sphere formed of so incompressible a substance as rock can shrink only by some sort of folding or displacement of its surface. Therefore, as the cooling of our globe proceeded, explosions and swellings constantly occurred at weak points or lines on or near the surface, where the prodigious strain forced a break. That these upheavals were most prominent and extended in the northern hemisphere is shown by the fact that the great masses and heights of land are grouped there; and the trend of mountain-ranges seems to show that the range of breakage and upheaval was in general in north-and-south lines. Elsewhere, and mainly in the southern hemisphere, broad areas of perhaps stiffer crust sank downward, making the vast depressions into which poured the waters of the primeval sea, and where our oceans still sway and roll.

All these changes, however, have been in the direction of insuring more

SURF AT FORT DUMPLING, R. I.

and more stability; and when the ocean water had thoroughly cooled, the very chill of its vast masses in the depths of the troughs assisted in the work, for the cold water, by more rapidly withdrawing their heat, caused the rocks beneath their basins to become denser, thicker, stronger, and consequently less liable to break or change, than were those rocks forming the foundations of the continents.

The moment it had shores to beat upon, that moment the ocean began to knock them to pieces under its pounding surf, and to grind the fragments so small that they could be drifted away, reassorted, and deposited wherever the water was sufficiently quiet to let them fall. The original rocks — chiefly granite — held the different forms of lime, magnesia, etc., to make the limestones; the silica to make the gritty sandstones; the alumina to make the clays; and so on. The sea not only was the agent to eat this old, rich crust to pieces and respread it into strata, but to sort out for us the materials to a considerable extent, laying down beds of limestone by themselves, and sandstone, shales, marl, etc., by themselves. It is probable, says Professor Shaler, that layers of rock twenty miles in thickness have thus been laid down on the gradually settling ocean floor, much of which has been raised again to form continental lands.

Hitherto we have spoken of the waters that surround the continents as if they formed one mass, as, practically, they do; but for convenience' sake we may designate certain areas by separate names, which ought now to be defined. Thus the larger, more open spaces are known as *oceans*, and of these five are recognized, namely, Pacific, Atlantic, Indian, Arctic, and Antarctic. Parts or branches of these, more or less inclosed by land and usually comparatively shallow, are termed *seas*.

The Pacific Ocean is the largest, it alone covering more space than all the continents combined, having a breadth, east and west, of ten thousand miles (about the length of the Atlantic), and an area of seventy million square miles. The equator divides it into the North and South Pacific. The former is comparatively free from islands, and is inclosed northward by the approaching extremities of Alaska and Siberia; while the latter widens at the south into the boundless Antarctic Ocean. Its basin is a vast depression of fairly uniform depth, studded in the western part by island peaks,—the summits of submerged volcanic mountain-ranges. The name "Pacific," or "Peaceful," was given to it by Magalhaens (Magellan), its first navigator, in 1540 (see Chapter IV), in his joy at having escaped from the tempestuous experience he had long endured in the South Atlantic. On the whole the Pacific deserves its name as compared with the Atlantic — a fact chiefly due to its great size. The term " South Sea " was

PERCÉ ROCK, IN THE GULF OF ST. LAWRENCE, SHOWING
DESTRUCTION OF SHORE-ROCKS BY WATER.

formerly much used for it, but English-speaking persons now usually mean
by that phrase the island-studded district between Hawaii and Australia.

The *Atlantic* commemorates in its name the myth of Atlas and his island.
Atlas seems to have been originally, among the Greeks, the name of the
Peak of Tenerife, of which they had vague information from the earlier

Phenician sea-wanderers. Then this was forgotten, and in place of the fact arose a myth of a Titan who stood upon a vast island in or beyond the "Western Sea," called Atlantis. Legends of wars with its people form a part of the nebulous hero-story of the beginnings of Athens; and it is said to have sunk out of sight long before records began. There have always been those who believed this story founded upon fact, and only a few years ago a book was printed in the United States arguing that the tale was the history of a real land; but not only is there no literary or historical evidence that Atlantis had any firmer foundation than vague memories of the Cape Verd or Canary Islands, but every evidence of the geological condition and history of the eastern shores and bed of the middle Atlantic Ocean shows that no such convulsion as the destruction of this island calls for ever took place there, or that there was ever such a land to be submerged. The Atlantic occupies a long, winding, comparatively narrow trough, that measures about ten thousand miles north and south, from the ice of the Antarctic to the ice of the Arctic ocean, and has only a few islets south of Iceland, the Faroes and the Shetlands, which rise from a plateau stretching from Labrador to Great Britain, the higher points of which were probably above the water within comparatively recent geological times, possibly since man appeared upon the globe. The average depth of the Atlantic south of this ridge is about thirteen thousand feet, but greater depths are found along the African and American coasts, on each side of a long submerged ridge from which rise the isolated islands of Cape Verd, St. Helena, and Tristan da Cunha. The width from Norway to Greenland is only about eight hundred miles, but between Montevideo and Cape Town it is thirty-six hundred miles, and the average width is about three thousand miles. The shape and situation of the Atlantic make it the most stormy of the three great oceans, and it is the one where the phenomena of tides, currents, etc., are most prominently manifested, as we shall see. It is also the most frequented and best known, because it has been necessary to study it for the benefit of commerce.

The Indian Ocean is simply the extension of the vast southern waterzone northward of parallel 40°, south latitude, where, from the Cape of Good Hope to Tasmania, it is six thousand miles in width. At this line the depth suddenly decreases, as though the edge of a submerged Antarctic plateau defined the southerly rim of its basin there. This ocean contains several large and some groups of small islands, but these are mostly near the shore, and connected with the neighboring continent by shallow waters, showing that they rise from a submerged plateau. The average depth of the Indian Ocean is about fourteen thousand feet; its surface-water is warmer and salter than that of any other; and its winds and weather are more regular and peaceful than in either the Atlantic or the North Pacific.

The Arctic Ocean is the well-defined body of water around and probably over the north pole. It is connected with the Pacific only by the narrow and very shallow Bering Strait, and with the Atlantic by comparatively narrow openings. It has been fairly well explored as far north as the parallel of 80°, and found to contain many islands; but it appears that there is great depth of water north of Spitzbergen and northeast of Greenland, making it probable that the trough of the Atlantic reaches to or beyond the pole itself. Most of its area is covered with drifting ice.

The Antarctic Ocean is regarded as the space of water within the Antarctic circle; but this is surrounded by a zone of deep ocean, unbroken almost half-way to the equator, except by the narrow southern part of South America and by New Zealand. It is an area, apparently rather shallow, of ice, fogs, and tempestuous gales, inclosing lands of unknown extent.

But these geographical distinctions are merely convenient methods of speech. After all, there is only one ocean "poured round all," and its particles are incessantly changed in place and remingled by means of a world-wide system of tides and currents, the effect of which is to keep sea-water everywhere uniform in character and perfectly pure and healthful.

WAVE-WORN CLIFFS AND PEBBLE-BEACH AT ÉTRETAT, FRANCE.
(FROM A PAINTING BY WILLIAM P. W. DANA.)

IN MID-OCEAN: A GREAT WAVE.

CHAPTER II

NOW that we have studied the ancient ocean, it is time to study its present characteristics and understand the great and important part it plays in the world.

A very striking thing about the ocean is its flatness. Being water, it seeks always to find its level; and we commonly assume that it everywhere does so, and take the sea-level as the standard from which to calculate all heights above or depths below its surface; that is, we assume that every part of the surface of the ocean when calm and at mean tide is exactly the same distance from the center of the globe. This, however, is not wholly true. Careful observation has shown that the Pacific is several feet lower on the western shore of the Isthmus of Darien than is the Atlantic on its eastern shore — a fact due, no doubt, to the crowding of water by the Gulf Stream into the Caribbean Sea. The Mediterranean is known to be somewhat higher than the Atlantic, and other differences exist in similar places elsewhere.

This introduces the subject of depth — a matter which we have learned accurately only within a very few years. In the early days ropes alone were used for sounding, and these had to be of considerable size to bear the strain; but a mile or so of rope became too heavy to handle, and depths below that length remained unmeasured. Then a little machine was tried consisting of a heavy weight having attached to it, by a trigger, a wooden float. This was thrown overboard. It sank, and when it touched bottom the shock released the float. From the time that elapsed before the float reappeared the depth was estimated. This, however, was little better than guesswork; and accurate soundings exceeding one thousand fathoms were not obtained until an American naval officer began to use wire instead of rope. From this hint was developed elaborate machinery, operated by steam, using steel piano-wire, having automatic registers of the amount reeled out, and carried down by weights that were released when the bottom was struck, making it easier to recover the wire. To these weights (or rather to the wire just above them) were attached devices for clutching and bringing to the surface specimens of the bottom, self-closing jars to fetch water from the lowest layer, self-registering thermometers that re-corded the temperatures at the greatest or at various intermediate depths, and other means of learning the character of the water, bottom-material, and animal life several miles below the surface, including methods of photographing by aid of a submerged electric light. Such investigations, carried on in ships suitably equipped, have been prosecuted by several governments, most notably by the expedition of the *Challenger*, a British surveying-ship which circumnavigated the globe during the years from 1872 to 1876.

SEA-CAVE NEAR GIANT'S CAUSEWAY, NORTH OF IRELAND.

This and many other expeditions have sounded in all parts of the world, and explored large tracts where the water uniformly exceeded three miles in depth. The United States ship *Enterprise*, after passing the Chatham Islands in her run from New Zealand to the Strait of Magellan, found the water everywhere more than thirteen thousand feet deep. Throughout her run from Montevideo to New York the water varied from twelve to eighteen thousand feet deep, and Captain Nares and Admiral Belknap found like depths over equally vast breadths elsewhere.

Yet even in these basins more profound pits and valleys exist. Several places are known near Japan and off Porto Rico exceeding five miles in depth; and an English officer sounded 29,400 feet in the southern Pacific Ocean, nineteen hundred miles east of Brisbane, without finding bottom.

The average depth of all the oceans is estimated at from twelve thousand to fifteen thousand feet. As, according to Humboldt, the average height of the lands of the globe is only about one thousand feet, it will be seen that all the land now above the water, and its foundations, could be shoveled into the ocean troughs and still leave water more than two miles in depth covering the whole planet.

The soundings and dredgings of which I have spoken enable us to make a tolerable map of the ocean beds and to describe their features. All the continents are bordered by a shelf reaching out under the shallow shore-water to a greater or less distance, and then dropping, usually with much abruptness, to the ocean trough. This shelf, perhaps originally a part of the primeval continent, bears most of the great islands near continents, such as Newfoundland, the West Indies, Great Britain and Ireland, Madagascar, the Aleutian, Japanese, and Philippine groups, the Malay Archipelago, and others. If you will look at a map that has marked upon it the line of one thousand fathoms' depth along the shores of the various continents, you will find it reaching far out from the eastern shores of both Americas, the western and northern shores of Europe, the eastern shores of South Africa, prolonging India hundreds of miles, and embracing great spaces among the East Indies, while even the hundred-fathom line would connect many an island with the mainland or with some other island, as they actually have been connected in times gone by. The fact is, there is not a single proper mountain-peak rising out of deep water at any great distance from the margins of the continents. All the numerous islands of the wide oceans are either coral reefs or the summits of volcanic cones.

Upon this shelf, and for the most part within two hundred miles of the coast, are deposited all of the materials torn from the land by the sea or brought down by rivers or glaciers, excepting the very finest, which currents may float somewhat farther out, and also excepting the rocks that icebergs carry away and drop in mid-ocean: but this is not a great amount, for most icebergs strand on the shallows off Newfoundland or in Bering Sea.

Almost nothing from the shores, therefore, reaches the central depths of the open oceans, whose beds are in substantially the same condition that

they were in at the beginning, except for two things — volcanic upheavals
in some places, and the remains of animal life everywhere. The former exception is a very important one, since it is now known, according to Professor Shaler, that volcanoes, by their eruptions, send more dust and broken
materials to the seas than the rivers and shores combined.

"Although the deeper sea-floors probably lack mountains," says Professor Shaler, "they are not without striking reliefs, which, if they could

THE VOLCANO KRAKATOA (SUNDA STRAIT) IN ERUPTION IN 1883.

be seen, would present all the dignity which their size gives to the Hima-
layas or Andes: the difference is that these elevations are not true mountains, but volcanic peaks, sometimes isolated, again accumulated in long,
narrow ridges, but all made up of matter poured out from the craters or
through great fissures in the crust. So numerous are these heaped masses
of lava and other ejections from these vents that there is hardly any considerable area of the oceans where they do not rise above the surface.
There are indeed thousands of these volcanic peaks distributed from pole
to pole. . . . Thus on the floor of the North Atlantic there is evidently a
long, irregular chain of these elevations extending from the Icelandic group
of islands southward to the Azores. If an explorer could view this part
of the sea-bottom, he would probably find that the line of craters was as
continuous as that exhibited by the volcanoes of the Andes.

"Besides the volcanic peaks," Professor Shaler continues, "the sea-bottom in certain parts of the tropics . . . is beset with the singular elevations formed by coral reefs." But of these I shall have more to say toward the end of the book, and I allude to them here only as a feature of the invisible landscape beneath the waves.

Over the vast, gently undulating spaces separating these submerged lines of volcanoes and the ridges of coral, lies a mat of mud of unknown thickness, which naturalists term "ooze." It is principally composed of volcanic dust and of the microscopic "tests," or flinty or limy skeletons, of minute animals, few of which are large enough to be seen by the unaided eye. "Dwelling in myriads in the superficial parts of the sea, these foraminifera, as they are termed, sink at death to the bottom, over which they accumulate a thick coating of minutely divided limestone powder, forming a layer of ooze as unsubstantial as the finest snow."

In regions like the North Atlantic this ooze consists almost wholly of such animal matter; but in other regions, such as the South Pacific, where volcanoes prevail, it is constantly and largely increased by an enormous quantity of mineral matter hurled broadcast by volcanoes, all of which are on islands or near sea-coasts. A part of this is the merest dust, which slowly settles from the air, perhaps hundreds of miles from where it was ejected. A larger part consists of that spongy lava called pumice, which is so full of holes filled with air and gases that it may float half way around the globe before it sinks, as happened after the explosion of Krakatoa.

Into the oceanic ooze, too, sinks so much of all dead fishes and other mid-sea animals as is not dissolved or devoured before reaching it; and it forms the grave of thousands of men. It is often said that ships and other things would not sink far, but would float, suspended by dense water or some miraculous influence, only a few hundred or a few thousand feet below the surface, for no one knows how long. But this eerie notion has no foundation in fact. "No other fate," we are assured by those who know, "awaits the drowned sailor or his ship than that which comes to the marine creatures who die on the bottom of the sea. In time their dust all passes into the great storehouse of the earth, even as those who receive burial on land." Wooden wrecks probably last much longer than those of iron.

I have mentioned that a small part of what the sea tears away from the land, or receives from rivers, winds, and other sources, is dissolved in its waters, which now contain, no doubt, samples of every ingredient of the rocks and soils of the dry land, and very likely some elements not yet detected. This solvent power of the sea explains its saltness, and it must go on growing more and more bitter as long as its waves grind at the shores and the rivers run down. The salinity varies in degree, water at great depths being salter than that near the surface, and excelling in saltness where evaporation is rapid, as under the trade-winds, while fresher in the regions of equatorial calms, where an immense amount of rain falls; broadly, the lightest (freshest) water is found at the equator, and the heaviest in the temperate regions. Inclosed, or nearly inclosed, areas become very salt. Thus the Dead Sea is what chemists call a saturated solution, being nearly one third (28 per cent.) salt, and Great Salt Lake in Utah is not far behind. The Red Sea contains 4 per cent., and some parts of the Mediterranean nearly as much. Taking all the open oceans together, about 3½ in every 100 parts (3½ per cent.) is composed of various salts, more than three quarters of which is common salt (chloride of sodium), and the remainder mainly forms of magnesium. One of the *Challenger* authors has estimated that the oceans contain enough salt to make a layer 170 feet thick over their whole area, and another writer says that the amount, if heaped up, would be four times larger than the whole bulk of Europe above the level of high-water mark, mountains and all.

In early times, indeed, sea-water, which yields about a quarter of a pound of crystallized salt per gallon, was almost the only source of salt for food. Even yet it is the principal source of supply for the manufacture of commercial salt in France, Portugal, Spain, Italy, Austria, the West Indies, and Central and South America; and it is largely used in Holland, Belgium, and Great Britain. The early process, still extensively practised in some parts of Europe, was to admit the sea-water to large partitioned flats floored with clay, where it evaporated rapidly. The salt-crystals remaining were then collected, purified to a greater or less degree, and sold off-hand. It was by similar means that our great-grandfathers in New England and along the Southern coasts provided themselves with salt, only they used large vats arranged over fires instead of earthen basins exposed to the sun.

But analysis of sea-water discloses small quantities of many other recognizable minerals. Silica must be there to supply the needs of many foraminifers, sponges, and other animals: lime in various forms exists, or else such sea animals as mollusks could not compose their shells, nor polyps erect their enormous reefs; bromine is present, and to the iodine and other mineral dyes in the water we owe the lovely purples, crimsons, and scarlets painting corallines, seaweeds, echinoderms, and some molluscan shells, as that of the Sargasso-snail (Janthina).

As for gold and silver, both are present. I have seen it stated that a voyage of a year or two is sufficient to permit the formation of a film of silver all over the copper sheathing of a ship's bottom, so that a frigate returning from a long cruise is really silver-plated; but I fancy this is more a matter of imagination than visible reality. Gold, in certain chemical combinations, certainly exists in sea-water, and may be extracted therefrom. Up to the present, however, the cost of the extraction has been more than the precious metal obtained was worth. Gold is often washed from sea-sand.

The ceaseless restlessness of the ocean forms another of the greatest contrasts between it and the immovable land — *terra firma*, as those like to call it who have been tossing too long on the "rolling deep." This characteristic restlessness involves some of the most important and interesting facts in physical geography; for were the waters still,— that is, were the oceans simply huge, quiet ponds,— none of that action could take place along the shores which has been so important an agent in shaping the world and making it a suitable place for human habitation and social development.

A FIORD, OR DEEP CREVICE WORN IN SEA-CLIFFS.

On a planet with an atmosphere and changing seasons like ours, however, a stagnant ocean is as impossible as a motionless air; indeed, it is because the air *is* always in motion that large bodies of water are never at rest, for it is the changing density and temperature and movements (winds) of the air that produce waves and currents.

Waves are caused by the pressure and friction of the wind upon the surface of the water, as you may readily see at any pond; and the water in them simply rises and falls, driving forward a little at the very surface so as to cause a gentle current called *wind-drift*. When the waves approach the shallow, sloping border of the land they are checked at the bottom by the slope of the beach, while the freer upper part goes forward, and the waves speedily lose their rounded form and become more and more sharply ridged and steep on the front side as they sweep on until at last they pitch forward in the crash and thunder of surf.

In the open ocean the waves are usually doing little work except to cause the surface to rise and fall. The harder the wind blows, the higher the waves become, and the faster they travel. This speed has been calculated, and has been found to be proportionate to size.

"Waves 200 feet long from hollow to hollow," we are told, "travel about 19 knots per hour; those of 400 feet in length make 27 knots; and those of 600 feet rush forward irresistibly at 32 knots." These, of course, are under the furious impulse of a gale, and it is marvelous that ships can be made to ride over them; nor is it any wonder that excited mariners clinging to the bulwarks of some small and heeling craft, should call them "mountain high," and declare in all seriousness that they have seen their crests rising one hundred feet above their hollows. No such altitude, nor half of it, probably, is ever reached by a storm-wave in the heaviest cyclone. An excellent authority, Lieutenant Qualtrough, assures us that the highest trustworthy measurements are from forty-four to forty-eight feet. The height of a wave depends upon what mariners call its "fetch"—that is, its distance from the place where the waves began to form. This has been worked out mathematically by Thomas Stevenson (father of the late Robert Louis Stevenson, the novelist), an eminent engineer and designer of lighthouses, who gives the following formula: "The height of the wave in feet is equal to $1\frac{1}{2}$ multiplied by the square root of the fetch in nautical miles." If the waves began 100 miles away from your ship, the waves about you will be 15 feet high, because the square root of 100 is 10, and one and a half times 10 is 15 (feet). The highest waves are not formed in the greatest tempests, which beat down their crests, but when the gale is both very strong and long continued. The worst "seas," as sailors call big waves, are those met with off the Cape of Good Hope and Cape Horn.

The depth to which wave disturbance extends depends on the violence of the wind, and near shore upon the slope of the bottom. Prestwich tells us that pebbles may sometimes be moved at the depth of one hundred feet, and sand much deeper, as is shown by the fact that the bottom is disturbed in heavy storms on the Banks of Newfoundland.

The weight and power of such on-rushing masses of water are tremendous, as appears from the effect on coasts where they strike; but this opens up a subject which is too large for treatment here, and I must refer readers to geological treatises, and to such special works as Professor N. S. Shaler's excellent "Sea and Land," where the work of the ocean in tearing down and building up its coasts is fully and entertainingly explained. I shall have

LOW TIDE, ST. JOHN'S HARBOR, N. B.

something more to say on this point, also, when I come to the chapter "Dangers of the Deep," and speak of the terrible destruction caused by earthquakes, and in certain other agitations of the sea not due to the wind, and often styled "tidal waves." There is only one kind of "tidal wave," properly speaking, however; and this is a theoretical rather than an actual one, perceptible usually only in that rising and falling of the water along coasts twice each twenty-four hours that we call the flow and ebb of the tides; and here we see the effect rather than the thing itself.

The tide has been an inevitable circumstance of the existence on the earth of the ocean, or any other great body of water, ever since its origin, yet it was not until Sir Isaac Newton made us comprehend the law of gravitation that its mystery was explained. We now know with certainty — if you want the mathematical formula and so forth, consult some good modern encyclopædia under the word *tide* — that this periodical rising and falling of the sea is due to the attraction of the sun and moon, — to the last three times as much as to the first, because it is so much nearer. This attraction is exerted toward the globe as a whole; and its visible effect upon the movable water is to lift it bodily on that side nearest the moon, and at the same time to pull away the earth from the water on the opposite side, which amounts to the same thing; and thus high tides are simultaneously produced at these antipodes, which accounts for the two a day. At the same time, however, the intermediate spaces have low tides caused by an attraction there toward the center of the earth. "There are thus always simultaneously and directly under the moon two high waters opposite each other, and two low waters at equal distances between them. Owing to the rotation of the earth, this permanent system of swells and troughs travels from east to west over every part of the ocean and of its coast, and explains the regular succession of rising and falling waters, at equal intervals of time, which we call the tides."

But the sun also exerts a similar but lesser influence, producing four daily solar tides, which most of the time are lost to view in the greater lunar tides. When, however, the moon gets into line with the earth and the sun, so that both the heavenly bodies pull together like a tandem team,

THE EARTHQUAKE-WAVE PASSING OVER THE LIGHTHOUSE ON POINT ANJER.

as happens twice a month,—at new moon and full moon,—their combined action causes unusually high water, which is the sum of the lunar and solar tides, and is called the spring tide. High water is then highest, and low water lowest. On the other hand, in the midst of these fortnightly intervals, when the moon is at its first or third quarter, the sun is a full quarter of the heavens (90°) away from the moon. Its influence, therefore, acts at right angles to or practically against that of the moon, and the solar tides go to swell the low waters and diminish the high waters, forming what sailors call neap tides,—preserving an old English word meaning *low*.

Now remember that the globe is not standing still, even while we make these explanations, but is revolving at a tremendous speed, so that the water under the moon lifted by lunar attraction is changing place every instant at the rate of over one thousand miles an hour, and you have the conception of a low wave on each side of the earth, reaching north and south, highest and swiftest on the equator and diminishing toward the poles. These are the true tidal waves. Were the globe covered with an unbroken mantle of water, such waves, each about twenty inches (or twenty-nine inches at springtide) high on the average at the equator, would follow one another round and round the earth at the rate of one complete circuit in every twenty-four hours. That must have been the case in the primeval ocean before any continents existed; and something of it still exists in the belt of unobstructed water surrounding the Antarctic continent of ice. It would then be flood tide or ebb tide at the same hour along the whole length of any one meridian. But in the present condition of the globe, where the oceans are separated by continents and broken by islands, the progress of the tidal waves is obstructed, deflected, and wholly stopped in a great variety of ways and places, so that the hours, amount, and behavior of the tides are exceedingly varied in different regions, and are often very puzzling, forming one of the most difficult matters with which the practical navigator has to deal. Interference of tidal currents forms the Maelstrom, off the coast of Norway, whose revolution is reversed twice daily, the classic Scylla and Charybdis, in the Straits of Messina, so much dreaded by the navigators of old, and many other whirlpools of less celebrity. The tidal wave sweeping northward across the Atlantic has time to round the northern end of Scotland and flood the German Ocean with southward swelling currents before the rising water pouring into the southern end of the English Channel has time to push its way through that narrow and shallow passage; hence the two floods meet in the Straits of Dover, which accounts for the miserable chop-sea so sadly prevalent in that unfortunate bit of water.

The natural height of the tide seems to be from two to five feet, as shown in the midst of the broad Pacific. "But when dashing against the land, and forced into deep gulfs and estuaries," to quote Professor Simon Newcomb, "the accumulating tide-waters sometimes reach a very great height. On the eastern coast of North America, which is directly in the path of the great Atlantic wave, the tide rises on an average from 9 to 12 feet. In the Bay of Fundy, which opens its bosom to receive the full wave, the tide, which at the entrance is 18 feet, rushes with great fury into that long and narrow channel, and swells to the enormous height of 60 feet, and even to

70 feet in the highest spring tides. In the Bristol Channel, on the coast of England, the spring tides rise to 40 feet, and swell to 50 in the English Channel at St. Malo on the coast of France."

To this cause is also due in some degree those great oceanic currents which form another striking fact in the history of the sea; but they are mainly due to temperature, wind, and the rotation of the earth.

The drops that make up a body of water are the most restless things in the world; they are always sliding down the least slope, sinking out of the way of lighter substances, rising to let a heavier object pass beneath them, or moving hither and thither in an ever hopeful search of that levelness and quiet that we call equilibrium. Furthermore, when water is heated it becomes lighter. Should, therefore, a portion of the sea grow warmer than the remainder, it must and will rise to the surface; and whenever a portion becomes cooled, it must and will sink.

Now, under the continuous blazing sun of the torrid zone the sea-water near the surface gets fairly warm,—having an average temperature of about 85° along the equator,—while in the polar regions the ocean is always chilled by permanent or floating ice until it is nearly cold enough to freeze; but these masses of warm and cold water cannot remain separate in the universal ocean. The hot tropical flood, continually rising, *must* flow away somewhere to find its level; and it can flow nowhere except toward the poles, for there the ever-sinking volume of chilled and therefore heavier water sucks it in to take its place, while it, in turn, creeps underneath toward the equator, there to fill the gap which the escaping warm water leaves behind. So we know there is constantly going on an interchange of water—a constant flowing *away* from the equator northward and southward on the surface, and a flowing in *toward* the equator along the bottom; an endless springing up in the torrid zone and a steady settling down in the polar seas. One out of many proofs of this fact is that the thalassal abysses below the depth of a mile or so are known to be ice-cold. This could not happen unless they were constantly filled and refilled with new water from the great coolers at the poles; for if the water at those depths should remain unchanged, it would soon become very warm from the heat of the interior of the earth, whence it does constantly extract some heat.

But while this invisible *vertical circulation* is going on, another more visible and interesting set of movements is in progress on the surface, forming what are known as *ocean currents*. These are vast rivers in the ocean flowing across its face in certain directions and to a certain depth, as rivers make their way along the land. They begin and are kept going mainly by a union of the two causes already explained — heat and wind.

The heat of the sun at the equator, warming, lightening, and evaporating the water, constantly tends to draw the colder water from the poles, most copiously from the South Pole; but the Antarctic water, hastening to the equator, is soon interrupted by the extremities of Australia, Africa, and South America, and so split into three great branches. That which passes into the South Atlantic goes on northward along the western coast of Africa, part of it becoming so warm under the hot sun there that it will not sink, but constantly comes more and more

A STEAMER BORNE ASHORE BY
AN EARTHQUAKE-WAVE.

to the surface, until it strikes against the great shoulder of Guinea and is turned sharply westward. Now it is squarely under the trade-wind and headed the same way; constantly urged forward by this moderate but endless tugging of the wind upon its waves, the current can never swerve, but flows along the equator, and for half a dozen degrees each side of it, straight across the Atlantic. South America, however, stands in its path, and the wedge-like coast of Brazil, pointed with Cape St. Roque, splits this great river. Part of it now turns southward and swings back across toward Africa, making an eddy a couple of thousand miles wide in the South Atlantic, while another arm runs down the Patagonian coast. But by far the largest part of the divided current is sent northward, past the coast of upper Brazil into the Caribbean Sea and Gulf of Mexico, where it is well heated, and thence poured into the North Atlantic, to become widely celebrated as the *Gulf Stream*.

Gathered in full force, the Gulf Stream flows northward close along the coast of our Southern States at the rate of eighty or ninety miles a day until Cape Hatteras gives it a swerve away, when it strikes out to sea and pushes straight across to Spain, where a branch leaves it and runs northward between Iceland and the British Islands, while the main body turns southward to mingle again with the equatorial current from Africa and repeat its journey all over again. It is in the heart of this great circle of currents in the middle of the Atlantic that navigators find that dreaded region of heat and calms which they call the Doldrums; and here, too, float round and round the wide, buoyant meadows of the Sargasso Sea.

Meanwhile another most important cold stream is making its way through the Atlantic, known as the Arctic current. It comes down out of Baffin's Bay, joins a similar flood from the outer coast of Greenland, is thrown up to the surface by the Banks of Newfoundland (where meeting warm air, it produces those thick and prolonged fogs so common in that region), fills the Gulf of St. Lawrence and the bight between Nova Scotia and Cape Cod with chilly water, and finally dips under the Gulf Stream amid that commotion of winds and waters that makes the track of the steamships between New York and Europe the most tempestuous of ocean highways. It is the mingling of these warm and cold waters there which is chiefly responsible for the stormy condition of the North Atlantic.

The Pacific has a similar arrangement of circulation north and south of the equator. The Antarctic waters form a cold stream named the Humboldt current, which pours up the western side of South America, keeping the climate down to a far more wintry condition than it is entitled to by latitude, until it reaches the southern trade-winds, which sweep it westward straight across the Pacific, where much of it is lost among the archipelagoes of Oceanica, and the southern part flows onward into the Indian Ocean.

North of the Pacific equator a similar westward current moves steadily over the great waste of waters past the Sandwich Islands to the coast of China. From the Philippines and Japan northward, however, there is a far stronger flow, known to the Japanese as the Black Current (Kuroshiwo), which skirts the coast of Japan and the Kurile Islands, makes these and Kamchatka habitable, then turns sharply east along the front of the foggy Aleutian chain of islands, and broadening and cooling as it turns, swings down the temperate coast of Alaska and gradually disappears. These two great currents and their inclosed eddies are far broader and less distinct than those of the North and South Atlantic, but they follow the same laws.

In a similar but lesser way the Indian Ocean has a strong westerly stream flowing straight across from Australia to South Africa, which is of immense help to ships returning from the East around the Cape of Good Hope. From Mozambique the water turns northward to make the return round, but here it is complicated by the peculiar conditions made by the inflow and outflow of the Red Sea, Arabian Gulf, and so on, and by the disturbing influences of the monsoons, until it can hardly be defined.

Of all these currents none is as well marked as the Gulf Stream. Its blue water is in such contrast to the darker, greenish hue of the remainder of the ocean that sailors can often tell when they enter the edge of the current, half their vessel being in and half out of the stream. If you approach from the west you find that the water at first shows a warmth of only fifty or sixty degrees near the surface; but as you sail on, this increases until, opposite Sandy Hook, you may get as high a reading on the thermometer as eighty degrees, and opposite Florida above one hundred degrees. This difference in temperature between the eastern and western margins of the Gulf Stream is owing to the presence of the great river of Arctic water flowing in an opposite direction between the Gulf Stream and the shore. Off Florida the Gulf Stream is about sixty miles wide; off New York it is over one hundred miles in width, but is less sharply defined. Its depth is hard to determine, but certainly amounts to several hundred feet. It is worth remembering that, although some guesses had been made at it before, Dr. Benjamin Franklin was the first man to study the Gulf Stream and to tell us anything of its origin and course.

The way in which some of these ocean currents affect the weather of the lands upon which they border shows how great is the influence of the sea upon land-climates; indeed, it may be truthfully said that only the continents and such great islands as Australia or Madagascar have any climate essentially distinct from that of the ocean in their quarter of the globe. But the equability that would reign over an ocean of quiet water, determining the amount of cold and heat by regular gradation in latitude between the equator and the poles, is completely upset by the great current-movements I have outlined. Scotland, for example, lies as far north as Labrador, and the latitude of London is above that of Lake Superior, yet neither have those terrible frosts and heavy snows which prevail in Canada, and make Labrador a land of ice almost uninhabitable. This difference is due almost wholly to the fact that the Gulf Stream pours its warm flood against the coast of Great Britain, and even tempers the Norwegian coast, keeps

Barentz's Sea largely free from summer ice, and clothes Spitzbergen with vegetation, although within ten degrees of the pole. Hence in the forests of northern Scandinavia Laps can dwell in much comfort on a line with the frozen barren grounds north of Hudson Bay.

On the other hand, the unfortunate coasts of Greenland are bathed in water chilled by months of captivity near the pole, and loaded with ice that cools down all the winds that blow ashore. Greenland itself is covered with an unbroken sheet of ice, hundreds or thousands of feet thick, yet most of it is no farther north than Sweden. The whole northeastern

A ROUGH NIGHT IN THE GULF STREAM.

coast of America, down to Labrador, is incrusted with ice; and the region south of the St. Lawrence has a similar climate to Finland; while even farther south, Boston, within the protecting arm of Cape Cod, is in winter a city of frost and snow and fog from November till April, when it really is little farther north than sunny Naples, where one laughs at winter.

Similarly, in the Pacific Ocean, the northward movement of the great Japanese current makes the coast of China habitable and pleasant clear to the Sea of Okhotsk, gives the Aleutian archipelago a pretty decent climate, and causes the islands and coasts of Alaska and British Columbia to nourish the most magnificent forests in America, and to have a climate resembling that of Great Britain. Glasgow and Sitka are, in fact, in the same latitude, and under very similar climatic conditions, except that in Scotland there are no such lofty and cold mountains to precipitate constant rains as is the case along the northwestern margin of America.

Similar examples and contrasts might be drawn in other parts of the world. The weather in the interior of continents is pretty much alike on similar latitudes the world round, varying with height; but the climate of all sea-coasts is good or bad as a place to live, in accordance with the temperature of the water which the currents bring to that part of the ocean.

But the currents of the ocean influence something besides the weather. Upon them depends to a considerable extent whether a certain part of the coast shall have one or another kind of animals dwelling in the salt water. This is not so much true of fishes as it is of the mollusks or "shell-fish," the worms that live in the mud of the tide-flats, the anemones, sea-urchins, starfish and little clinging people of the wet rocks, and of the jellyfishes, great and small, that swim about in the open sea.

Nothing would injure most of these "small fry" more than a change in the water making it a few degrees colder or warmer than they were accustomed to. Since the constant circulation of the currents keeps the ocean water in all its parts almost precisely of the same density, and food seems about as likely to abound in one district as another, naturalists have concluded that it is temperature which decides the extent of coast or of sea-area where any one kind of invertebrate animal will be found. It thus happens that the life of Cuban waters is different from that of our Carolina coast; and that, again, largely separate from what you will see off New York; while Cape Cod seems to run out as a partition between the shore life south of it and a very different set of shells, sand-worms, and so forth, characteristic of the colder waters to the northward.

Out in the ocean, however, the warm current of the Gulf Stream forms a genial pathway along which southern swimming animals, like the wondrously beautiful Portuguese-man-o'-war (Physalia), may wander northward for hundreds of miles beyond where they are found near shore; yet if by chance they stray outside the limits of the warm Gulf Stream, they will at once be chilled to death, as happened once to millions of tile-fish.

Ocean currents carry floating burdens long distances. They bring the icebergs to form those terrible fogs of Alaska and Newfoundland; and they often bear far away the logs that float out of tropical rivers.

These drifting logs often have plants growing upon them or contain quantities of seeds which are not injured by their short voyages. When, therefore, the coral polyps build up one of their reef-islands until it appears above the waves, thither the currents bring roots and seeds from neighboring islands, and quickly plant them upon the new barren shores, so that in a few seasons the little islet becomes green and wooded and ready to hold its own against the winds and waves. Moreover, the same drifting stuff will carry land animals as passengers,—insects, snails of many kinds, reptiles, and even four-footed beasts,—and so not only give the island a vegetation, but populate it with various of the smaller animals. This seems to you, perhaps, a very accidental and haphazard way of fitting out a country so that presently it may support human beings, nor is it the only means by which barren islands become productive; but it is important as far as it goes, and when we study into the distribution of plants and animals in an archipelago, we are pretty sure to find those of the same sort upon islands that lie in the same current—even to the human inhabitants.

A YOUNG SHIP-RIGGER.

CHAPTER III

THE BUILDING AND RIGGING OF SHIPS

S late as 1861 an exploring ship was visited by natives of Western Australia, riding simple rough logs. To smooth and sharpen the log's end and then to hollow it out has been thought to be the first step taken by primitive man in his progress toward a boat; but I think the dugout probably came later, or at any rate no earlier, than the folding of bark into a trough and tying up the ends, as some savages are still content to do. In North America, where materials were favorable, this germ developed into the very highest type of canoe — the Algonkin birch-bark. It may have been an attempt to imitate the bark canoe in a more durable form which led to the laborious hollowing of dugouts; but here again, in regions where suitable trees grew, the art developed so highly as to produce the great sea-boats of the Papuans and our Northwest Coast Indians, carved from a single log, yet able to carry sixty or more persons and their luggage. Such boats as these, when provided with sails, are practically "ships," and satisfy every need of their owners.

Another root of naval architecture lies in the raft, which long ago reached a high degree of usefulness in the sea-going balsa of western South America. It is probable that the South Sea catamaran is a clever outgrowth of experience with a raft. In Polynesia it took the form of two great canoes, exactly equal, fastened close together and covered by a single central deck; and such are the seaworthiness and speed of these double boats, that the Polynesians voyage hundreds of miles in them.

Similar in purpose—namely, to insure stability—are the various outriggers that at once characterize and distinguish among themselves the native craft of the South Seas. This device consists of a beam of the lightest obtainable wood, usually about half as long as the canoe, which rests upon the water parallel to and a few feet away from the side of the boat, and is connected with its gunwale by elastic rods or planks. Sometimes these are covered, or partly covered, by a light platform, and there are many variations in form: but the idea in all cases is to keep the boat from overturning.

In many parts of the world logs could be obtained large enough only for a narrow bottom or hollowed keel, and the remainder of the boat was built up of planks and pieces ingeniously pegged and knit together with treenails, ratan, and cords made of vegetable fibers that tightened when wet. The Madras surf-boats are a familiar example in civilized waters of boats made in this way which have great elasticity, and out of them have developed, without much change, the swift proas of the Malays, and the junks of China, Korea, and Japan. One device for stitching these boats firmly together was the leaving of ridges on the inner side of the planks or pieces, through holes in which they could be tied to each other and to the inner framework without making a hole reaching the outside. This system seems to have been earlier than the use of treenails.

PROA, WITH OUTRIGGER.

Of similar construction, apparently, were the boats of the Egyptians and other peoples about the eastern end of the Mediterranean and the Red Sea, which, as far back as three thousand years before Christ, at least, had reached the size and capabilities of true ships, making, as we shall presently see, extensive sea voyages. Pictures of them remain in the very ancient tombs, and show that the planking consisted of pieces about three feet square, which were laid on overlapping, like shingles on a roof, and fastened to the framework by wooden treenails. The Phenicians, and their pupils the Greeks and Romans, improved on these methods in various ways, at last substituting iron, copper, and bronze nails or bolts (which would not rust) for the wooden pegs of their ancestors.

All of these boats and those of all western Europe (of which the best outside the Mediterranean were the vikings' ships) differed in one essential point of construction from Oriental ships: instead of making the shell of the vessel, and fitting into it a framework of connected braces, as the Malays and Polynesians did (and yet do), they laid a keel, bending it up or setting into it stem- and stern-posts at the ends, and inserted along its sides curving upright timbers, well styled "ribs," which swelled out amidships, and narrowed in forward and aft, making a skeleton of the shape the hull was intended to be. Finally, over and upon this well-braced framework were securely fastened the planks, which were narrow and ran lengthwise

in every case except that of the ancient Nile boats. The Scandinavian
vikings developed a craft of their own, one of the most interesting of the
ancient ships; and to these northern craftsmen is traceable the principal influence that has shaped British (and consequently American) ship-building
and seamanship. This early Scandinavian boat was always made of oak,
sharp at both ends, and rather shallow, the general form being much like
that of a modern whaleboat, with a great rounding keel — if, indeed, this
wonderful sea-craft may not be a lineal descendant of the viking ship.
The hewn planks were attached to the keel and to the ribs (usually single,
naturally bent V-shaped prongs of oak) in a most ingenious and serviceable
manner, and they were always overlapping or clinker (i. e., clencher) built.
Several of these and other prehistoric boats have been found buried in peat-
moss and in mounds in Germany, Denmark, and Scandinavia, and have
been described by various writers.

The motive power of all the early boats was found in human arms,
wielding paddles or oars. It is said that the oldest forms of paddles of
which we have any record among the Egyptian or Assyrian hieroglyphs
show them to have been shaped somewhat like the arm and hand, and that
similar paddles were to be seen a few decades ago on the canals in Holland.
This is natural, because undoubtedly the first paddle ever used was the
naked hand. Short paddles were soon found less powerful than long ones;
but in order to work the latter it was necessary to brace them against something in the middle. Notches were therefore cut in the edge of the boat, or
thole-pins were inserted, the paddle became an oar, and by and by boatmen
learned the art of feathering, and so forth.

Steering could be done of old, as now, with a turn of the rearmost paddle in a canoe, and as canoes enlarged, the steering-paddle was lengthened.
As the sterns of the ancient boats were usually either sharp, like the prows,
or else built up into an ornamental height, the most convenient place for
the steering-oar was over the right side, where it was balanced in a loop of
cable, or otherwise, as close to the after end of the boat as practicable, and
then a cross-piece extended inboard from the handle, enabling the steersman to move it more easily by giving him the benefit of leverage. Such
was the arrangement of steering-gear in all the ancient Mediterranean
boats, and it is to a similar arrangement in the sea-going craft of our northern ancestors that we owe our words *stern* and *starboard*, which originally
meant "steering-place" and "steering-side." The modern rudder is substantially the same oar, set upright, tiller and all, and hinged to the stern-
post; in fact, the word has descended from the old Teutonic name for "oar,"
and all gradations between steering-oar and true rudder may still be found.

Though some romantic stories are told by the old mythologists as to its origin, the idea of rigging was as natural and practical in its development as that of hull or steering-gear. That a strong breeze moves a canoe, and that, if a man in a canoe holds his robe outstretched or a thick bush upright, the force will send him along without the labor of paddling, and lengthwise rather than sidewise, because that is the direction of least resistance, were facts quickly and gratefully seized upon by the earliest boatmen. To have a skin ready for the purpose, and to set up a pole and ropes to hold it in position, were easy matters; yet in this simple arrangement you have the first sail.

But skins were too heavy and valuable for such a purpose, except in such limited circumstances as those of the Arctic Eskimos.

Persons who spent much time on the water, therefore, like the most ancient Egyptians and the islanders of the Chinese and South seas, soon devised a way of weaving rushes or splints of bamboo into broad mats, and thus were able, on account of their lightness, to carry much larger and more effective sails, which were kept outstretched by one or more cross-poles or spars, and could be taken down quickly. Many such sails are in use to this day not only among Asiatic and African boatmen, but on the northwest coast of Canada. A fine example hangs above my desk as I write.

With the discovery of how to make cloth and cordage of woolen, silken, hempen, and cotton fibers (and in Egypt of papyrus), came a still better material for ropes and sails, since cloth was so much lighter that a far greater extent of it could be spread than before; its flexibility enabled it to be handled, changed, and rolled up snugly, and its cheapness encouraged its use and the practice of navigation generally. We read of silken sails on the royal barges of medieval times, but they could hardly have exceeded in strength or elegance those of the fine Phenician ships that carried the commerce of the world twenty-five centuries ago. "Fine linen with broidered work from Egypt was that which thou spreadest forth to be thy sail," exclaims the sacred chronicler (Ezekiel xxvii. 7). Hempen cloth, indeed, was preferred for sails until the present century, as is expressed in our word *canvas*, which is derived from the Latin name of flax; but now cotton has mainly superseded it.

Anciently the sails were often colored, purple or vermilion being the badge of a monarch or an admiral. Black denoted mourning. "In some cases the topsail seems to have been colored, while the sail below was plain; and frequently a patchwork of colors was produced by using different stuffs." Various inscriptions and devices were also woven or painted on the sails, sometimes in gold. The Venetians and Greeks do the same

REEFING A TOPSAIL IN A STORM

to this day, adding a gaudy feature to the lovely Levantine sea-scenery; and the sails of the North Sea fishermen are turned to a rich red and yellow by the tanning mixture in which they soak their canvas.

As for the shape, all rigs seem reducible to two types — the lateen and the square. The former is characteristic of the eastern half of the world, the latter of the western half, including primitive America, where, so far as I know, only plain, rectangular sails were ever made by the Indians.

There must be some good reason for a broad division like this, and it is found in the different conditions which eastern and western seamen had to meet. The lateen seems to have originated in the Indian Ocean, is seen wherever Arabs are, and has been taken eastward by the Malays as far into the South Sea Islands as their influence extended. It is a huge, triangular canvas extended at a steep angle by a long, flexible yard balanced across the mast to which it is loosely hung, and controlled by a sheet attached to the free corner. It is thus very lofty, and therefore suitable to a region of steady and usually light winds. This is the characteristic rig of the Arab dhow — a model that has come down from remote antiquity and is capable of excellent service on the northern and eastern coasts of Africa, where it prevails. It was probably in a small vessel of this kind that the Apostle Paul suffered shipwreck; and an outgrowth and perfection of it is the dahabiyeh of the Nile, now become famous as a tourists' pleasure-boat,

A HONG-KONG "PULL-AWAY" BOAT.
Showing method of hoisting and reefing matting sails.

whose immensely lofty sail is precisely adapted to catch every faint breath that comes across the river from the deserts. Such sails are spread like the great pointed wings of an albatross over the narrow decks of the Malayan "flying proas" and other swift South Sea craft, and urge upon their fleet errands the xebecs, saics, feluccas, and other light craft of the Levant and Barbary coasts, identified with former piracy and modern smuggling, as well as with fishing and freighting. Some of these boats have two or three masts, the xebec and felucca being notable because of the curious forward rake of the foremast; and in that extremely picturesque Portuguese fishing-boat called the muleta there are, in addition to the big lateen, a huge free second sail ballooning out to leeward from the tip of the yard, and a host of little flying jibs forward, which somebody has well likened to a flock of birds hovering about the prow. Good examples of lateen-rigged boats may be seen in Louisiana, built and manned by the Greek, Maltese, and Sicilian fishermen.

The difficulty of handling in rough or squally weather this long yard and expansive canvas makes it unsuitable for such weather as prevails in

the western Mediterranean or on the Atlantic; and to meet these stormy
and frequently changing conditions, and obtain a rig with which they could
beat to windward, the earliest rough-water seamen devised square sails.
What the rig of the ancient far-voyaging Phenician ships was we have no
means of knowing, but the indications are that they carried lug sails, which
appear to be the simplest and earliest of the "square" forms; that is, sails
suspended from short cross-yards, and controlled by ropes (sheets) attached
to their lower corners.   Such at least were the sails of the Roman and
Greek merchant and war vessels of the classical era, and they persist to-
day in the local fishing-smacks of the stormy Adriatic.

The true home of the square-sailed craft, however, was northern Europe,
where the Norwegian, Dutch, and Norman coasters and fishermen of to
day probably represent fairly well the rigs of the bold viking boats of
twelve or fifteen centuries ago.

Of the slow development of ship-building during the middle ages we
have little information, but in the fourteenth century we begin to hear of a
revival in the art, as, indeed, was needful when the long voyages were to be
undertaken which the discovery of the mariner's compass had then rendered
possible.   In this revival the Venetians and Genoese took the lead, but the
English were not far behind.   There was a large variety of vessels in that
day, rude though they were, and called by names we should hardly recognize.

Though the hulls of these vessels were large and tight, their shape was
poorly adapted for speed or for safety in bad weather.   Their decks were built
up into immensely high structures at the stern and bows, after the old galley
model, and to form forts for soldiers.   Our word "forecastle" reminds us
of this old usage.   Their masts were single sticks,— not divided into topmasts,— and hence, necessarily, were thick and heavy; and they bore upon
their summits large "top-castles" where marines stood in battle to shoot
down upon the enemy's decks.   This weight above, with the height of
surface exposed to the wind and the clumsy rigging, made it impossible
for them to sail safely except with a fair and gentle wind (they never attempted it otherwise), and they were required to carry an enormous quantity of ballast.   There was so little room for anything except armament,
sleeping-berths, and a cooking-room in the war-ships that every war fleet
had to take with it small vessels carrying provisions; and the case was
little better in respect to merchant vessels.

The ships in which Vasco da Gama, Columbus, the Cabots, and other
explorers did their marvelous work were no better than this.   Strangely
inefficient they seem to us, and we wonder that some of the simplest contrivances in rigging were not adopted centuries before they came into use

3

until we remember that it was not for long, speedy voyages that vessels were intended previous to the sixteenth century (with certain exceptions in northern seas), but simply as a means of carrying slowly from one coast-port to another a great number of men or huge cargoes.

However, as the known world widened and trade grew, inventions by private ship-owners continually improved the rigging, though it would be hard to find a class of men slower to change old ways for new than the sea-men. Columbus's "caravel" had four short masts, the forward one having a square lug-sail and the three after masts lateens. It was very gradually, indeed, that lateens were given up, and most curious combinations of sails were to be seen in this transition period of the fifteenth and sixteenth centuries. The old-fashioned Mediterranean barca, for example, had as foremast the forward-raking "trinchetto" of the felucca, with a huge lateen, while the mainmast bore three square sails and the mizzen two lugs; and in addition to this two banks of oars were provided! In fact, it was not until 1800 that English frigates substituted a spanker for the lateen-rigged mizzen.

Another curiosity of rigging possessed by these solidly built, beautifully carved vessels (no such exterior decoration has been seen since as adorned the ships of the sixteenth and seventeenth centuries) was the quaint little spritsail-topmast. By this time the single heavy pole-mast had been super-seded by the three built-up masts and topmasts, braced by stays, made accessible by rope ladders (shrouds), and carrying several tiers of topsails instead of only one. A bowsprit had been added, also, and this became almost a fourth mast, so loaded were it and its stays with various small sails. Its outer end bore this miniature spritsail-mast, with topmast, shrouds, and tiny sails all complete, surmounted by a pole-head, or jack-staff, upon which was hoisted the flag since known as the jack, and always now carried at the prow of any national boat or ship, even such as the shapeless monitors.

But gradually, out of the experience of long voyages, the competition of merchants, and as an effect of improved gunnery and consequent changes in naval tactics, the lofty deck-structures, great tops, needless outworks, and odd sails, like this spritsail, were got rid of, and vessels were trimmed down and equalized until they became, as now, "ship-shape, Bristol-fashion."

The rigging of modern sailing-vessels is divided into "standing" and "running"; the former includes the masts, their stays, now generally made of wire, and such other rope-work as is not adjustable.

The sails, also, may be assigned to two classes: first, those attached to a mast, with or without boom and gaff, or to a stay, which are called fore-and-aft sails because they may be ranged lengthwise of the ship; and, second, those suspended by their upper and lower edges to or between spars or "yards" swung across the mast, and known as "square" sails, the lowermost of which are really lugs. All the variations of shape seen in America, except the rare and local lateens, can be counted in one or the other of these classes.

The styles of rig visible in American waters are not many, and are easily learned. Let us begin with the simplest — that having one mast.

The *cat-boat* (*i. e.*, cat-rigged boat) is one having a simple pole-mast stepped very near the bow, and a fore-and-aft sail laced to a gaff and boom

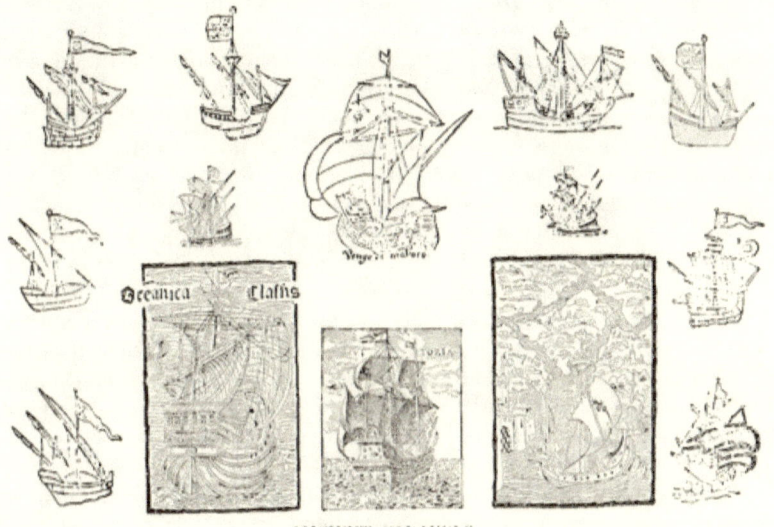

ANCIENT CARAVELS.
Copied from old manuscripts and tapestries.

and managed by a sheet. This is the rig of the ordinary American sail-boat, which is noted for its ability in pointing up into the wind. In England it is known as a una-boat. Sometimes the peak of the sail is sustained by a little loose spar called a "sprit," instead of a gaff. In the chapter on Yachting will be found further illustrations of these small rigs.

A *sloop* has one mast (with topmast) set well back from the stem, and a bowsprit. The sloop-rig consists of a fore-and-aft mainsail, spread by means of a boom and gaff, a gaff-topsail, a forestaysail, and one or more jibs. A *cutter* is now substantially the same thing, though formerly somewhat distinguished. Both are derived, probably, from the northern lugger, and old-time pictures show queer intermediate forms, often having a square topsail instead of a gaff. Thus the earlier of the Hudson River

sloops, which were not only the freight-carriers but the packet-boats between New York and Albany from the time the Dutch introduced them
until steamboats took their place, had the top of the mainsail supported,
lug-fashion, by a short yard, and carried above that a square topsail; but
this rig was steadily modified toward the modern type to make it faster
and safer in the sudden squalls that beset this hill-girt river.

Of two-masted rigs, the oldest is the *brig*, which has square sails on
both masts, just like the main and mizzen masts of a full-rigged ship.
Then there is the *brigantine*, a slight modification of the brig, and the *her-
maphrodite brig*, or *brig-schooner*, with fore-and-aft sails on the after mast.
This kind of vessel has been greatly modified (one of its most extraordinary
forms was the *ketch*), is less common now than formerly, and took its
name, which is derived from the same source as "brigand," from the fact
that it was the most common rig of the pirates of the sixteenth and seventeenth centuries. Its place was largely taken for small vessels by a purely
American invention, and one of the greatest of Yankee notions — the
schooner. The schooner was originally small, and had two masts; but now
is often built of great size, with as many as five or six masts, each of which
has a fore-and-aft rig — that is, a sloop's mainsail and gaff-topsail on every
mast, with forestaysail and several jibs in front, and staysails between.
Sometimes a square sail is placed on the foretopmast, which makes the
vessel a *topsail schooner*. The first one was built by a Gloucester sea-captain about 1817, and proved so satisfactory that all the fishing-fleet were
soon rigged in that way, whence the idea has spread to all parts of the
world.

Until recently, however, vessels large enough to have three masts were
always "square-rigged," as *barks, barkentines*, or *ships;* for, although we
have come to speak of any big vessel as a "ship," yet in proper nautical
language a ship is a vessel rigged in a particular way, and it is nothing
else. In fact, in olden times they were sometimes very small — too small
to be economical, as we now know. The "Naval Chronicle" for 1807 contained an account of a full-rigged ship of only thirty-six tons' burden,
which for one hundred and thirty years previous to that date had been
cruising about the English coast, and may be doing so yet, for aught I know.

Masts have their proper names: the tallest is in the middle of the vessel,
and is called the *mainmast;* the next tallest stands in front of it, and is the
foremast; and the third is in the stern, and is named *mizzenmast*, because
it carries the mizzen (sail). All the rigging, except that belonging to the
bowsprit, is repeated for each mast, and each piece is named with reference
to the mast or part of the mast or appropriate sail to which it belongs: as,

for example, main shrouds, fore shrouds, mizzen shrouds, mizzen-royal, main-topsail yard, foretopmast studdingsail downhaul, and so on. In a proper full-rigged ship all the sails upon the masts, except the spanker, are square, and are named from the sections of the mast opposite which they hang. Counting from the deck to the truck, or tiptop of the mast, they are as follows: on the mainmast, mainsail or maincourse, maintopsail, maintopgallant-sail, main-royal, and skysail; on the foremast, foresail or forecourse, foretopsail, fore-

A FIJI ISLAND OUT-RIGGED CANOE, APPROACHING A FULL-RIGGED SHIP HOVE-TO.

topgallant, foreroyal, and skysail; on the mizzenmast, cross-jack (and behind it the spanker, mizzen, or driver), mizzentopsail, mizzentopgallant, mizzen-royal, and skysail. The bowsprit sails are the forestaysail, foretopmast staysail, jib, flying jib, and outer jib, or jibstaysail. Each of the stays running diagonally from mast to mast bears a triangular sail known by the name of the particular stay on which it hangs, as maintopmast staysail, and so on—nine in all. In addition to all this, a little sail is sometimes set above the skysail, and another under the bowsprit, while out beyond the ends of the yards are often extended light additional spars carrying stud-dingsails. In favorable weather, when the captain wishes to "crowd all on," as sometimes can be done for days and weeks together before the trades, almost forty sails may be spread, and the ship moves grandly along under a

3*

swaying cloud of canvas that reaches far beyond her rails on each side, and towers more than one hundred feet into the steady air.

But the cost of building, maintaining, and handling these grand fabrics is so great that they are steadily diminishing in numbers, and perhaps are destined before long to disappear altogether from the seas to which they have lent so much picturesqueness and romance. The supremacy of the schooner seems likely to prove complete. Unwilling to concede everything at once, many vessels are now rigged with square sails on the foremast and mainmast and fore-and-aft sails on the mizzen (a *bark*), or square sails on the foremast only, and the others schooner-rigged (a *barkentine*); but even these are disappearing in favor of the three-masted or four-masted schooner. This is due to the fact that the schooner rig will sail closer to the wind and gives as much force in proportion as the ship style, while it is far less expensive to build, and more quickly and easily managed, not requiring nearly as many men, and therefore being cheaper to run as well as to set up. It is for these reasons that I have called it one of the greatest of Yankee notions.

A MULETA, OR PORTUGUESE LATEEN-RIGGED FISHING-BOAT.

CHAPTER IV

EARLY VOYAGES AND EXPLORATIONS

PART I — PREVIOUS TO THE DISCOVERY OF AMERICA

HEREVER it may have been that man first appeared upon the earth, the period must certainly have been incalculably long ago, for he had time to spread to all parts of the habitable globe long before any sort of record begins. Little, if any, part of the world has yet been found where the evidences of man's residence in the long-forgotten past do not exist. So long ago that all tradition of it is forgotten, and only the imperishable stone implements they used remain as traces of their presence, mankind had reached and settled the farthest northern and eastern coasts of Europe and Asia, and the southern extremities of Africa and India. These might have been reached by land; but similar traces exist in many islands which, so far as we can see, could never have been connected with each other or with any continent by lands now submerged (as perhaps has been the case in some other islands) since man originated. Such places, then, could have been reached and colonized only by means of boats, and that at an exceedingly remote time.

Some hint of what these prehistoric navigators might have been able to do may be gathered from the performances that we know of in the South Sea, where almost every island and coral atoll that could support a colony has apparently been inhabited, since long before even tradition begins, although some of them, like the Hawaiian group, are separated from all others by hundreds of miles of open sea.

It is exceedingly interesting and suggestive to read in a work like Professor Friedrich Ratzel's "History of Mankind," of the dispersion of population over the islands of the South Pacific Ocean, where a mixed population of black and yellow races possessed themselves of the whole of Oceanica long before white men had even heard of that part of the world. This astounding fact gains in significance when we remember that wide tracts of very deep ocean divide these islands, many of which are so small that they were found by exploring navigators only with difficulty. Cook and Beechey and other early voyagers note finding upon certain islands people who had come thither in their own boats over distances of six or eight hundred miles; and there are many instances of castaways surviving voyages of one thousand or fifteen hundred miles, even against the trade-winds. But these involuntary voyages were no longer than many others undertaken for war or trade, or because of famine or a mere love of wandering. Over-population of the limited spaces of most islands and groups led to the colonization of others; and it must often have been necessary to go far away to seek unoccupied or thinly peopled refuges. This could not have been done had men not been good shipwrights, not only, but careful students of the heavens by whose sun and moon and stars they steered, aiding themselves with charts made of sticks. The remotest groups, like the Sandwich Islands and Easter Island, were found and set-tled too long ago even for tradition to retain more than a fabulous story about it. "These Vikings of the Pacific," says Ratzel, "continued to dis-cover even small and remote islets. In the whole of the Pacific there is not one island of any size of which it was left to Europeans to demonstrate the habitability." It has even been argued that the continent of America was peopled by Pacific Islanders, who made their way to it from Polynesia; but of this there is no direct evidence, and it seems unlikely, because the prevailing winds and currents flow from South America, rather than toward it, in this part of the Pacific.

But leaving these dim old times when barbarous men voyaged far and wide over seas, and races mingled that were born on opposite sides of wide waters, let us note what traveling our civilized ancestors did.

The evidences of ruined walls, graves, carvings, and stone tools show that that earliest of civilized races of which we now have any knowledge — the Hittites — were acquainted not only with the coasts of the Mediterra-nean Sea, but had boldly rounded the headlands of Spain, skirted the stormy Bay of Biscay, and settled colonies in England and France. Who were these Hittites? They were an Asiatic people, dwelling in the Taurus Mountains of the eastern part of Asia Minor, who increased into the most powerful nation of that part of the world about two thousand years before Christ, and carried on wars with the Egyptians, among others, until at last they were overcome by the rise of the empire of Assyria, north of them, about eleven hundred years before Christ. Doubtless they explored the African coast somewhat south of the Red Sea, and very likely knew the

Persian Gulf and the route to India. My own opinion is that we are likely to give the people of antiquity too little credit rather than too much in the direction of a knowledge of geography.

Meanwhile there was rising along the Mediterranean from Palestine northward the most able commercial race of antiquity, who styled themselves Canaanites, as in the Bible, but whom the Greeks called Phenicians, the name by which we know them best. Their capitals were the cities Tyre and Sidon, the ruins of which are still to be seen on the Syrian coast a little way south of Beirut, and the wealth and commercial power of which will give us some interesting paragraphs for a future chapter. Suffice it here to say that their rulers were foremost among the loosely organized "nations" between the Nile and the Euphrates, and that they maintained their power through a long period, not only by their wealth and enterprise as traders, but mainly through their skill and energy as navigators. As we shall see when we come to consider their commerce in Chapter VII, they excelled in the building of ships, in an understanding of how to steer long courses by the heavenly bodies, and in sea knowledge generally. It is well known that the Phenicians traded in their ships down the west coast of Africa to and beyond the Canary Islands, which they also visited; made repeated voyages to the French coast and the British Islands; and may very likely have gone around into the Baltic, for they knew of its amber, though this might have been obtained by the overland trade routes. It is believed that they ascertained that Africa was, in fact, a huge island; for it was to prove this supposition that Pharaoh Necho (or Naku or Neku) II, an enlightened Egyptian monarch who reigned in the sixth century before Christ, hired a crew of Phenician seamen to man an expedition whose purpose it was to circumnavigate Africa. These men started down the Red Sea in 611 B. C., and in 605 B. C. came sailing home through the Strait of Gibraltar, to the delight of their friends and confusion of a kingdom full of I-told-you-sos.[1] Just twenty centuries elapsed before any one else repeated that feat, so far as I know; and no wonder it was forgotten. This same Necho II did even more for maritime commerce, for he attempted to complete the canal, begun long before his time, connecting the Mediterranean with the Red Sea, and seems to have made a passage along which barges and small boats might be towed, which remained open for many centuries.

[1] This is related by the Greek historian Herodotus, and has often been denied, especially by the older writers; but the "Encyclopædia Britannica" gives it credence, and tells us that the latest and best critic of the geography of Herodotus, Major Rennel, maintains the possibility of such a voyage, and believes it was made. He argues that the construction of their ships, with flat bottoms and low masts, enabled these hardy voyagers to keep close to the land, and to enter all the rivers and harbors for food and water. I think, therefore, that we may believe that Herodotus recorded what really happened, even if we reject some details.

and in part followed the line now covered by the Suez Canal. Earlier than that Darius, the Persian conqueror of Egypt, had dug a navigable canal from the Nile to the Red Sea; and this shows that there must have been large traffic in both seas at that time to justify such tasks.

By this time the power and prosperity of Tyre and Sidon had declined, and Carthage, originally a colonial city, had become the most important center of Phenician influence; and from this port there sailed a century later (perhaps about 500 B. C.) an exploring expedition under a Carthaginian king named Hanno, intended to study and establish trade with the West African coast. It was a large and powerful fleet, said to number sixty galleys; and that women were taken as well as men shows that it was intended to form settlements at suitable points, as, indeed, was done. The account of it has been preserved in a short writing called the " Periplus," by an ancient but unknown Greek; and this inscription is regarded by most scholars as entirely authentic, since all its details conform to modern knowledge, even though it is impossible to identify surely the various points mentioned. It tells us that the terminus of Hanno's exploration was an island beyond a gulf called Noti Cornu, in which he found a company of hairy women, whom the interpreters called *gorillas*. It was in memory of this that the manlike apes which a few years ago were discovered on the west coast of Africa received the same name; but they are not known anywhere north of the Kamerun Mountains, while the farthest point any critic is willing to believe reached by Hanno is the Bight of Benin, some

AN EARLY ROMAN BIREME.

distance north of the Kameruns. It is easy to believe that the inquiring Carthaginians might have heard of these apes,—or perhaps of chimpanzees, now found as far north as the Gambia River,—and reported actually seeing them, in order to add glory to their name. At any rate, this expedition increased largely the ancient knowledge of the sea in that direction; and navigators now knew the shores of the Atlantic from the Gulf of Guinea to the North Sea; but there the knowledge of the world seems to have rested for more than a dozen centuries, principally, no doubt, because there seemed nothing beyond, either north or south, to invite the merchants who then, as ever since, have been the principal promoters of discovery. It is only within the past century that voyages of discovery have been undertaken purely for the sake of the increase of knowledge. Previous to that the object was always either military conquest or the extension of trade.

Attention was now turned to the eastern seas, overland routes to India and even to China having become well known both to conquering armies and to mercantile caravans. The coasts of Abyssinia, of Arabia, of the Persian Gulf, and of western India were settled by a semi-civilized people for a thousand, perhaps two thousand, years before the Christian era; but they were broken into many independent tribes; and their ships, if they had any, only crept from one harbor to another near by, and neither knew nor cared what lay beyond the farther headlands. As time went on, however, and strong kingdoms arose in Egypt, Arabia, Syria, and Persia, consolidating these scattered tribes into nations, it became necessary to

SHIP OF PTOLEMY PHILOPATOR.

(About 240 B. C. Banks of oars and rig s. &c.)

learn the sea-routes between more distant ports. Thus it came about that while the Pharaohs still flourished, Arabic commerce extended regularly along the coast of Abyssinia, and doubtless as far southward as Zanzibar, while the Malays had probably already reached and colonized Madagascar. There seems no reason to doubt that those remarkable ruins in stone which the late Mr. Thomas Bent has studied at and near Zimbabwe, in Mashonaland, East Africa, are the work of Arabian gold-miners, made perhaps a thousand or more years ago; and it is pretty certain that Arabic seamen even at that date regularly traded as far as the island of Madagascar.

The Persian Gulf has been another nursery of a seafaring people since long before the record of history begins; yet so slow were they to learn of anything outside their capes, that it was accounted a wonderful thing when, in the winter of 325-4 B. C., Nearchus, the admiral of the fleet of Alexander the Great, voyaged from the mouth of the Indus to the head of the Persian Gulf. Soon afterward, however, under the house of the Ptolemies, rulers of Egypt, fleets sailed regularly between Red Sea ports and India and Ceylon.

But now for many long centuries the boundaries of the known world were not to be much enlarged (although methods of navigation were improved and commerce continued within the limits of Roman and Arabic dominion), for we know of the discovery of no new coasts until we begin to hear of the doings of an independent and far northern people, scarcely known to the civilized world, and certainly not regarded as a part of it.

On the bleak shores of the North Sea, where the fiords and creek-mouths of Scandinavia gave shelter not only from foreign enemies, but from each other, there had grown up a seafaring race of men, of Gothic ancestry, who had settled on the coasts of what are now Norway, Sweden, and Denmark. They styled themselves Norsemen, or men of the North, and did not object to the title Vikings, or Fiord-men; but their enemies called them pirates, and with much reason, for they ravaged and ruled all the coasts both north and south of the Baltic, voyaging northward to the "land of the midnight sun," colonizing northern France in the tenth century, and taking practical possession of all they pleased of the British Isles — Ireland and northern Scotland in particular. Here these Norsemen met equally fierce foes, or found congenial partners, as the case might be, in the Scottish and Irish seamen of that day, who were themselves bold freebooters and wide voyagers; and when, in the middle of the ninth century, the Northmen had discovered, as they supposed, the Faroe Islands and Iceland, a little exploration soon showed them that the Irish *culdees*, or priests of the Christian church planted in Ireland by St. Patrick, had been there before them — first in 725, according to the Irish chronicles of Dicuilus, who seems worthy of credence. Indeed, it is believed by some antiquarians that these Irish sea-wanderers had colonized Iceland at the same early age; had reached Newfoundland, and regularly resorted to its banks for fishing and whaling (five hundred years before Cabot); and were even acquainted with the coast of the North American continent, where traditions assert that their colonies were planted on what are now the shores of Virginia and the Carolinas, which they called New Ireland.

These are entertaining old stories, and may have some truth in them, for it seems certain that the Irish reached Iceland, at least, in the eighth century. Icelandic history, however, begins with the visits of Norsemen in 850, followed by others, who, a few years later, took colonies there and set up an island population which before a century had elapsed numbered more than fifty thousand people. They had a republican form of government, and were quite independent of the King of Norway (Harold the Fair-haired, great-great-grandfather of William the Conqueror), from whom the earlier colonists had fled because of his oppression; but they kept up acquaintance with the mother-country, and merchants and adventurers were continually voyaging between Iceland and all the islands and coasts of that region, using stanch vessels sometimes one hundred feet in length, and

eminently seaworthy; yet their only guides were the stars and such signs
as seafaring men read in the water and weather about them.

It continually happened, however, that they were driven far out of their
courses, in such a region of gales, currents, and fogs as is the North Atlantic.
In one such adventure, in the year 876, a sea captain named Gunnbjörn
Ulfkragesson was driven far to the west of Iceland, and when he got back
to port told his friends that he had seen land. Probably he also told them

A WAR EXPEDITION OF THE VIKINGS.
Showing build, steering-oar, and rig (ordered in a mid), of Scandinavian exploring ships in the North Atlantic.

that so far as he could see there was nothing but icy mountains, of which
they already had enough, for no one seems to have investigated the matter
further until more than a century later, when a turbulent viking of the
rebellious house of Erik, called Erik the Red, was banished from Norway
and fled to Iceland with his followers. He was soon convicted there also of
manslaughter in a neighborhood quarrel, and again condemned to banish-
ment. Iceland wanted to get rid of him and his brawlers, and Europe
would not let him return. Whither should he go?

Then his thoughts turned toward the strange land in the west that
tradition said Gunnbjörn had sighted. It is believed by the most careful

students that Gunnbjörn's "rocks" were volcanic islets, which have now disappeared, and are represented only by certain shoals; but it would not be incredible that he had caught a glimpse of the Greenland coast itself.

At any rate, Erik had little hesitation in starting out to rediscover them. Why should he? Those rough-riders of the sea were used to voyages of equal length. It is about 200 miles from the Norwegian coast at Bergen to the Shetland Islands; 200 miles from the Shetlands, or 225 from the Hebrides, to the Faroes; and 275 miles thence to the nearest coast of Iceland,—reckoning all in straight lines, shorter than any ship could actually follow.

If his viking boat and viking crew could span those stretches of sea unguided, what hindered his crossing the little further space whose tempests had no terrors for this wild sea-king? In that unpossessed land, could he find it, he might be free to riot at his will (but one cannot help thinking there was more in the man than that!); and if he could open to his people a new country, what wealth and power might not come with it to him, for the humbling of his rivals at the court of Norway.

So Red Erik sailed away to the west in 984, and two years later returned to Iceland and reported that he had met first a far-extending icy coast, along whose front he had sailed southward until he could turn to the west and then northward, thus rounding its narrow southern extremity (Cape Farewell); and there he had found a habitable region, which he called Greenland, in order, as he said, to attract settlers by a pleasant name. Thus this wicked old Norseman was the first of American "real-estate boomers."

Attracted by his story, a band of adventurers went back with him in 986, and established a settlement near the site of the present Danish town Julianshaab, just inside the cape, on an inlet that they named Eriksfiord.

Among these emigrants was one named Herjulf, whose son Bjarne [1] was a merchant captain who owned his own ship, and was then absent in Norway. Returning to Iceland shortly after Erik's departure, he concluded at once to follow his father, and, with a willing crew and still loaded ship, set sail for the west. But incessant bad weather drove them they knew not whither during many days. At last the wind fell, the sun shone out, and they saw land; but its appearance did not agree with the description of Greenland, and knowing they were too far south, Bjarne turned north, and kept on, occasionally sighting the coast, until finally he reached Eriksfiord in safety. No one knows what headlands he looked upon; but if the Icelandic versified chronicles called *sagas* may be believed,— and the wisest students of history put faith in them,— he was the first European to see America of whom we have definite knowledge.

[1] This is not a Norse, but an Irish name, familiar to us as *Barney*.

Several years passed by, however, before any one tried to profit by this accident and seek the lands that had been seen southward. Then Leif, the eldest son of old Red Erik, resolved to do so. He had talked with Bjarne and his men until he knew all the details of their story, and then he bought the same good old ship, and enlisted a crew of thirty-five men. This happened in Norway, where Leif then was, and it is said by some that the king aided and authorized the expedition. At any rate, after a public farewell they sailed away, and seem to have gone straight across the ocean; but whether they did this, or sailed by way of Iceland and Greenland, they easily found the unknown coasts Bjarne had described, and landed in Helluland, Markland, and Vinland, in the last of which they built huts and spent the winter of the year 1000.

The identification of these places has caused much discussion. That " Helluland " was Newfoundland and " Markland " Nova Scotia seems tolerably certain ; but historians are not agreed as to where that winter was spent in " Vinland," so called (meaning " Wineland ") because a German member of the crew gathered grapes there, from which wine could be made. When, in 1602, Gosnold discovered a fruitful island south of Cape Cod, he named it Martha's Vineyard, believing that he had found the place.

When Leif reached Greenland again, the next spring, every one was vastly interested in his discoveries, and emigrants from Greenland, Iceland, and even from Europe went out to colonize the new lands ; but the attempts, though spasmodically continued for a long time, seem never to have been really successful, so that no undisputed trace of the presence of these sea-wanderers on the mainland of North America is known to exist. That they knew the coast fairly well from Disco Island (70° N lat.) southward to Virginia, is generally believed ; but where Leif Erikson spent that first winter, or where the Vinland settlement of subsequent times was, is thus far a matter of conjecture. Some students of the sagas place it in New York harbor, others in Narragansett or Buzzard's bay, near Boston, or in Nova Scotia. Formerly the general belief was that Newport, R. I., or the shore above there, was surely the site ; but this was based, first, on the supposed European inscriptions on a rock in the Somerset River, at Dighton, just above Fall River, which were in reality only Indian markings ; and, second, upon the " old round tower " at Newport, which few persons now believe was built prior to the coming of the English colonists with Roger Williams. The late Professor E. H. Horsford believed that he had found the site of the principal Norse settlement of the tenth century, called Norumbega, at Watertown, on the Charles River, a few miles west of Boston ; and he made an argument from old maps, etc., to support his assertion that the ancient river-walls, etc., there were really the remains of a town; but historians generally do not attach any importance to Professor Horsford's theory.

Perhaps we shall never know where this "Vinland" was that Leif discovered, and where the queenly Gudrid dwelt and her son Schnorr—the first white child in America—was born; nor is it of much consequence that we should, for the settlements were few and transitory. That they existed, however, and that the shores of Canada and New England were occasionally visited from the tenth to the fourteenth centuries by Norsemen, cannot be gainsaid. That the Greenlanders did not all migrate to the warmer, well-timbered, and fruitful region in the south was probably due to the fact that

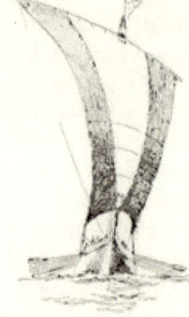

it was so remote from their kindred, and so open to attack by the native red men, whom they called *skrellings*.

Over the long but slow history of these American settlements of the Northmen we need not linger. Although Vinland seems to have been abandoned within a few decades, the Greenland settlements were maintained. A republican government was organized; Christianity was introduced, and remains of their stone churches and Augustinian monasteries have been identified. By the end of the fifteenth century, however, these colonies had completely disappeared, worn out in the hopeless struggle against climate and the savage Eskimos, but exterminated, at last, perhaps, by the Black Death—

A VIKING GALLEY.

for the great plague which almost depopulated Europe in the fourteenth century seems to have reached even the desolate shores of Greenland, and to have consumed the last of these remote people, causing them to be utterly forgotten.

A more definite account of pre-Columbian North America than that of the sagas and other traditions of the Vinlanders, and one accepted as true by Mr. Major of the English Hakluyt Society and other competent geographical critics, is that of the voyages and reports of the brothers Nicolò and Antonio Zeno. These men belonged to a family distinguished in Venice; and toward the close of the fourteenth century they separately or together made many voyages in the North Atlantic, going far beyond any previous navigators of which they knew. They wrote letters home containing an account of these, but little publication was given to them, and they were forgotten until the revival of interest in geography following the early discoveries of Columbus. The documents possessed by the Zeno family were then made the basis of a pamphlet by a grand-nephew reciting what his ancestor had done, long before the time of Columbus. The most interesting thing in it is an account of how, about 1390, Nicolò Zeno fitted out a ship at the Faroes, went over to Greenland and there learned of an island which was called Estotiland, and which we know as Newfoundland. Not very far away to the southwest of it, he says, was the country of Drogeo, which fishermen whom he saw had visited. They claimed to have "discovered" none of these places, but spoke of them as formerly well known, although then little frequented by Europeans.

As to Drogeo,— which he speaks of as if it were the mainland,— that was still occasionally resorted to for fishing; and he relates the adventures of a white man who had been captured by the mainland savages a few years previously, and adopted by them on account of his knowledge of how to fish with a net, and to do other useful things. Such a course would be very characteristic of the aborigines of eastern North America, as we have since learned to know; and it is also natural that he should have been fought for by rival chiefs, as Zeno says happened to this man, who, by capture and exchange, or of his own motion, traveled about and saw much of the people of this "country" Drogeo. At any rate, the information given by Zeno tallies remarkably well with the truth about primitive North America and its inhabitants. "They have no kind of metal," reported this wandering refugee, who finally drifted back to the coast, and was able to make his escape to a fishing-boat. Now the one really remarkable and distinctive fact about the North Americans was just this,— that with a considerable advance in other directions, they had never learned to fuse and forge or otherwise utilize iron or other metals, save a little metallic copper and silver in the Great Lakes region. But listen to the rest of his brief report:

They live by hunting, and carry lances of wood sharpened at the point. They have bows, the strings of which are made of beasts' skins. They are very fierce, and have deadly fights amongst each other, and eat one another's flesh [as was true, to a limited ceremonial extent, after battles]. They have chieftains and certain laws among themselves, but differing in the different tribes. The farther you go southwestward, however, the more refinement you meet with, because the climate is more temperate, and, accordingly, there [i. e., in Mexico] they have cities and temples dedicated to their idols, in which they sacrifice men and afterwards eat them. In those parts they have some knowledge of gold and silver.

Now, whether all this was the observation of a single rude sailor, or, as is more likely, summarizes what Zeno was able to learn from all sources at his command regarding the new western mainland and its people, it is correct and forcible. Had young Nicolò the editor, a century afterward, tried to invent something of the kind, he would surely have made his invention marvelous, for that was an age of fable and bombast. On the contrary, this is a simple and accurate statement of what we now know were the facts. Nor did he have any means of knowing anything more of the case than his family archives revealed, since he wrote and published this account of his uncle's voyages only a few years after the first return of Columbus, and before any writer had visited the northern American coasts, or had learned the habits of the natives. I can but believe, therefore, that the report was made in good faith, and records simply what the Zeni did and saw and heard; and that these bold Venetian navigators knew more about North America, at least, before the end of the fourteenth century than Columbus had learned by the end of the fifteenth.

I have run ahead of my story, but I wanted to show how little impression these northern investigations and occupation of a new continent had made upon the Mediterranean "world," which seems rarely to have heard of them, much less to have profited by the information, for more than four hundred years, in spite of the fact that there was constant communication between the Normans and British, at least, and the Mediterranean peoples.

Let us now go back to those southern countries and see what they had been doing toward maritime exploration during these thousand years and more when the Scandinavians were so busy in the north. It was principally perfecting the knowledge of the world their fathers knew. From the very first men had tried to make maps, and succeeded fairly well for small spaces; but to make a map of the whole world was a task that defied human knowledge for many centuries. After Aristotle's time all men of education understood that the world was a sphere; and about 150 B. C. Hipparchus, borrowing an idea from the Babylonians, taught the Greeks that the way to place their towns and mountains and rivers and the outlines of the coast correctly upon a model of the world, was to determine their position by observations of the heavenly bodies. Thus the ideas of latitude and longitude originated. He could not apply his method practically very far, because there were few or no accurate astronomical observations away from a few cities in Egypt and Greece; but two hundred and fifty years later Ptolemy, a learned mathematician of Alexandria, gathered all the facts obtainable, and made an attempt which bore a rude resemblance to the truth and served as the best and almost the only account of the world for several hundred years. Ptolemy flourished about 150 A. D. His book describes Asia as far east as the Malayan peninsula, Africa south to Zanzibar and the Gulf of Guinea, and shows a knowledge of Europe as far north as the Shetland Islands (Ultima Thule) and Denmark: the original work seems to have contained no maps, but these were added to it about 500 A. D. by another mathematician named Agathodæmon. It is called the Almagest.

Nothing of value was added to this during the long stagnant period of the world called the middle ages, when the love of learning declined and men fell back into the old traditions, even to the extent of being taught by their priests that it was a sin to believe that the world was round. In those times the Arabs of Bagdad nourished knowledge more than any one else, but even they did little for geography. Finally the people of Europe began to wake up and look at things for themselves, instead of tamely accepting whatever the Pope of Rome or somebody else told them, and going and

"Off, thou Norseland Terror, clouding   Which, the gods of old enshrouding,
Hellas with the jealous wraith   Froze their hearts, the poet saith!"

coming as he directed, regardless of whether it was for their interest to do so or not. One of the first and one of the most important influences of this revival in a desire for learning and the means for larger activity among men was the sudden extension of navigation; and this could not have come about, nor amounted to much, had the mariner's compass not been invented.

Nothing is more obscure than the history of this instrument. The Chinese have certainly known, from a remote antiquity, that a magnetized needle, permitted to move freely, would turn north and south; but they seem to have profited as little by it as by so many other useful things that, long afterward, in the hands of the more energetic men of the West, contributed so largely to the progress of civilization. They were accustomed to poise a sliver of magnetized steel upon a bit of cork and set it afloat in a bowl of water. One end was marked, but this, with characteristic Chinese perver-

sity, was the one pointing toward the south, not toward the north, as with
us. This rude and simple arrangement is still in use among the Koreans —
or was until recently. With such a contrivance and little, if any, knowledge
of the variation of the needle, the Chinese of a thousand years ago made
longer voyages than they have done in more modern times, trading not only
with India, but sailing regularly as early at least as the ninth century
to the Red Sea and the Persian Gulf.

There is no direct evidence, but it seems incontestable, that it was from
these eastern mariners that the Arabs received the compass, and gradually brought it into use in their home waters, where it became well known
to the crusaders and other sea-going travelers of the middle ages. Little
reliance could be placed upon it, however, until the sixteenth century,
when the need for something trustworthy for long voyages made men turn
their attention to the study and betterment of it.

Toward the end of the fourteenth century, as I have said, Europe was
beginning to recover from the terrible visitations of the plague, and to
wake from its lethargy and to look abroad; and various influences were at
work to promote exploration by sea and land — and what a grand field for
study there was!

At this time nearly all the commerce of Europe, mainly in Italian hands,
was with India and China. The overland route was long, perilous, and
expensive, and that across the Arabian Gulf hardly less so. At best, such
traffic was slow and limited, and the first need of the reviving world was the
discovery of some straighter and quicker road to the East. In this quest
Portugal came forward under the brilliant leadership of Dom Henrique
(Prince Henry), styled "the Navigator," who was the younger son of King
João (or John) I, and half an Englishman, since his mother was Philippa
of Lancaster. It was Prince Henry's ambition to extend geographical discovery and improve seamanship, and he enlisted the help of the best navigators obtainable, regardless of nationality. In order to observe the
heavens to better advantage, and also to study the tides and other nautical
phenomena, he established an observatory on the bleak headland of Cape
Sagres, where he willingly spent a large portion of his time for the sake of
science. Navigation was sorely in need of such help. Except that they
had rude compasses, of whose laws of variation, etc., they were ignorant,
the seamen of that day were little, if any, better equipped than were those
who sailed the "ships of Tarshish" a thousand years before that. Astronomers had supplied them with rough tables of the declination of the sun,
pole-star, etc., by which, with the help of a cross-staff, — a simple instrument for ascertaining angles, — they might make a guess at the latitude.

Longitude was found only by observations of eclipses of the moon, and noting the difference between the time when it was due at home, according to the almanac, and the local time of its actual coming; but at sea the "observations" were little better than guessing.

Chart-making was an important branch of study at Sagres. So few and rare were sea-maps then that one was never seen in England until 1489. To the collection of information in this direction, and the improvement of nautical methods, Prince Henry and his aids applied themselves most diligently; but he died before much had been accomplished. Nautical studies went on, however, under the next king, John II, for whom Martin of Bohemia, the foremost astronomer of his time, devised a form of the astrolabe for use on shipboard, increasing accuracy in finding latitude.

It was with no better instruments than these (and sand-glasses in place of chronometers) as guides over chartless and unsounded seas that the way was found to India and to America, and the globe was circumnavigated; and that the same thing might be done again is shown by the fact that only last year (1897) a vessel, which had barely escaped destruction in a storm and lost all her instruments in the mid-Pacific, was brought safely into San Francisco by observation of the stars and "dead-reckoning" alone.

But Prince Henry (for I have run ahead of my story again) was not content to study and teach on land alone. He was fired with the ardor of discovery and conquest likely to augment Portugal's wealth and influence in the East. Expedition after expedition was sent southward, and in 1435 Henry's ships finally passed Cape Bojador. Great was the wonder and rejoicing thereat, for it had always been taught by the monks that this cape was the end of the earth; but it was not until 1462 that the Cape Verd Islands and Sierra Leone were reached. Prince Henry had been dead since 1460, but the influence of his wise and untiring enthusiasm and work lived on, and inspired the king and people of Portugal to renewed efforts at solving that riddle of Africa that perhaps the Egyptian sphinx was meant to typify. By 1469 trade had been opened with the Gold Coast, and a few years later the mouth of the Congo was found.

These advances showed that there was nothing unnatural or fearful in the southern latitudes, as sailors had been taught to believe from time immemorial,—a superstitious dread which the old chart-makers long sustained by their habit of filling the empty sea-spaces on their maps with fearsome and wondrous monsters,—and therefore, in 1486, King John II sent Bartholomew Dias in two sail-boats — pinnaces of fifty tons each — with orders to go as far as he could; and this bold captain, passing the last known headland of the Guinea coast, sailed on and on, tracing the West

African coast, and landing here and there to examine the swampy shores, to get fresh water, and to hoist the castellated banner of Portugal in token of possession before the wondering eyes of naked negroes. At length he was blown and buffeted for days and days in heavy storms, and at their close found himself far to the eastward of his former longitude, whereupon he fought his way on and sighted land which he rightly determined must be the southern extremity of Africa. This was in 1487. Returning to Lisbon toward Christmas of that year, he reported his experiences, and

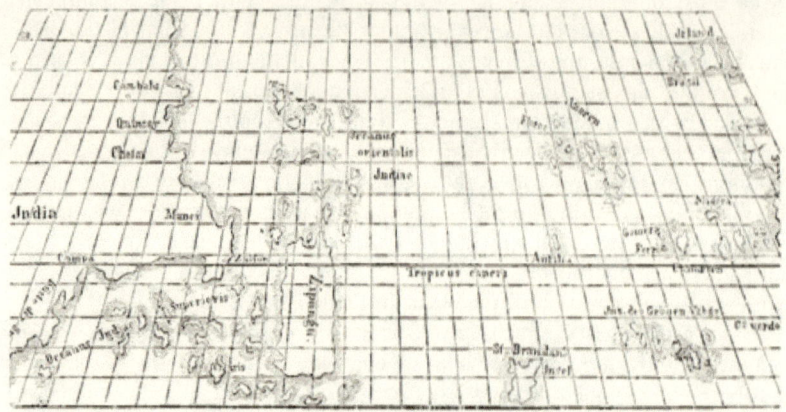

PORTION OF A FIFTEENTH CENTURY SEA-CHART, BY TOSCANELLI.

Copied by permission of Messrs. Houghton, Mifflin & Co., from Justin Winsor's "Narrative and Critical History of America"

dwelling especially upon the rough time he had had in the south, proposed to style the point of the continent Cape of all the Storms; but King John, foreseeing great things to follow for his country, said, "No; we will call it the Cape of Good Hope"; and so it remains to this day — but all the storms remain about it, too!

Now for some years previous to this time the monarchs of western Europe were much exercised over rumors of the existence somewhere in the Orient of an all-powerful and generally marvelous potentate styled (by them) Prester John, and reputed to be a conqueror of Asiatic, or perhaps African, infidels who later had become cut off from Christendom. The whole affair was a myth, probably arising from an indistinct knowledge of Abyssinia, whose negus afterward borrowed the title; but before this was realized popes and various "Catholic majesties" had sent embassies in search of Prester John's court, some of which incidentally gained valu-

able information. Among the latter was Pedro Covilho, an emissary of
Portugal, who, having failed to find Prester John in western India or Persia, made his way back to Egypt and Abyssinia, whence he sent home in
1486 or 1487 a report of progress that told John II some surprising news
of the advancement of the Arabs of that part of the world in the sci-
ences, and especially in those belonging to geography and navigation.

Covilho's messenger was a Portuguese Jew, Rabbi Joseph of Lamego,
who carried voluminous letters, one of which showed that Arabic mariners
were then familiar with the whole length of the east coast of Africa, including Madagascar, and were perfectly well aware where it terminated at
the south, and that there was no obstacle to passing around to the western
side of the continent; and just at this interesting juncture Dias came sailing
back in his pinnace to say that it was all true, for he had seen it.

Thus the sea-road was open to India and Cathay, and Portugal was
eager to take advantage of it. She was then one of the leading powers of
Europe, and the foremost one in colonial and commercial enterprise, striving to wrest from Genoa and Venice the supremacy in trade that they had
so long enjoyed. Nevertheless almost ten years elapsed before the next
expedition was sent southward to confirm Portugal's possessions, and establish commerce with the Orient. John II had died, and Emmanuel the
Fortunate reigned in his stead — a reign that has been called the heroic
period of the nation's history; and it must not be forgotten that "Little
Portugal" was then so mighty that a year or so previously (May 4, 1493)
the Pope (Alexander VI) had issued a bull in which he had divided, with
intended equality, all undiscovered parts of the earth between Spain and
Portugal, the former being given everything to the west, while to Portugal
were reserved all future rights east of a certain north-and south line.

The line of separation designated was the meridian of no variation of
the compass-needle. The existence of such a line had been discovered
by the same Christopher Columbus who was to thrill the world a few
years later; but he did not know, what only experience developed, that this
meridian was changeable, swinging many degrees east and then returning west in the course of two or three centuries. At that time the line
seemed fixed some three hundred miles west of the Azores, and philoso-
phers accounted for it later by a theory that it lay in the middle of the
Atlantic because there it was subject to an equality of attraction toward
both continents which held it steady. This was not true, but it was better
than the less learned but more popular explanation of the magnetism of the
compass — namely, that it was "an effluvium from the root of the tail of
the Little Bear." A year later, however (June 7, 1494), the treaty of Tor-

"THE SEA-ROAD TO INDIA AND CATHAY."

desillas, between Spain and Portugal, declared that the line of demarcation should be the meridian 370 leagues west of the Cape Verd Islands, or as nearly as possible in the center of the Atlantic. The supposition that there might be valuable lands within, that is, east of, that limit, inspired several of Portugal's subsequent searchers.

In 1497 King Emmanuel's expedition was ready to sail — the largest and best equipped, probably, that had ever been sent out by any government, and its commander was Vasco da Gama, a young naval officer of renown. His fleet consisted of four vessels,— small caravels, of course, one of which was commanded by Dias,— and left the Tagus, after ceremonious farewells, in July. Da Gama stopped at various places, but reached and safely rounded the stormy cape in November. He had with him the information (and some say an Arabic map) sent home by Covilho, but his business was not to verify this, but to reach India and establish new Portuguese possessions. Why, then, did he not strike straight across from Cape Agulhas, as East Indiamen have done ever since? For the good reason that he had no guide, no means of finding his way across the southern ocean, where all the stars were strange; for sun observations for latitude were then unknown to European navigators, and rarely used on land. Instead of this, he was obliged to turn northward and skirt the coast for a thousand miles, stopping here and there, until he had passed far enough north of the equator to bring above the horizon the familiar home stars, for which he had "tables."

At last, from the Arab port of Melindi, near Mombasa, he turned east

and sailed straight away to India, where he anchored before Calicut, then
the most important port of southern India, on May 20. Returning the
next year with ships richly laden, he was received with public rejoicings and
given high honors; and he greatly astonished his friends of the navy by
telling them that the Arabs used the compass, sea-charts, quadrants, and
" had divers maritime mysteries not short of the Portugals."

Da Gama lived many years, and sailed often to India and China after
that; but chiefly on political expeditions, in which he disgraced his otherwise great name by inexcusable rapine and cruelty.

Meanwhile some exploration had been done toward the far north, as we
shall see in the next chapter; and so the fifteenth century ended, with
Europe understood as far as Nova Zembla, Africa circumnavigated, and the
coasts of India, Malaya, southern China, and the larger Malayan islands
fairly familiar to geographers. This is much, and yet it leaves unmen-
tioned the greatest fact of all — the work of that grand, sad character, Christopher Columbus, upon whose grave near Seville has been written:

HE GAVE A NEW WORLD TO SPAIN.

THE FLYING DUTCHMAN.

" There, beyond the Cape of Storms,   " Men catch glimpses of the sail,
   Where the breaker's voice of thunder   Ages old, and rent and hoary,
   Roars when ships are rent asunder,   Of that quaint old ship of story,
   Through a fog of ghostly forms   And cry, ' Vanderdecken, hail!'"

THE ROCK IN THE SEA.

CHAPTER IV

(Continued)

EARLY VOYAGES AND EXPLORATIONS

PART II—FROM COLUMBUS TO COOK

HY to Spain? It is an "oft-told tale," and the merest reminder is all that is needed here. Columbus was a young seafaring man, born at Genoa about 1434, and ambitious to become a master of his profession, and especially to acquire great wealth. He traveled to Venice, Barcelona, and other cities where learning was to be gained, and became thoroughly acquainted with all the astronomical and geographical science of the time, and especially proficient in the art of cartography. Attracted by the naval activity in Portugal under that indefatigable Prince Henry, Columbus went to Lisbon about 1454, and endeavored to find a leading place in the sea-work that country was doing. But Portugal's eyes were so blinded by the glamour of Africa and the East Indies that she had no time to follow the gaze of this young and ardent Genoese captain whose eyes were turned steadily toward the west, where, more and more insistently, he urged that a sea-track, straight as a line of latitude marked on a globe, lay open to the Indies and the coasts of Cathay. To prove this true would be not only a glorious exploit for any man, but an achievement of untold advantage to the nation under whose flag he sailed.

Just how this conviction arose in the mind of Columbus we do not know. It was probably first a purely scientific conclusion from the facts of astronomy and geography that he had learned, encouraged by romantic traditions of western "Isles of the Blest." A few scientific men agreed with him, but the great influence of the Church of Rome condemned such notions as opposed to the Bible and revealed religion; and the mass of the people, ignorant and superstitious, looked upon them as foolish, and laughed at Columbus as a

PORTRAIT-STATUE OF COLUMBUS IN MADRID.

dreamer or worse. Between his danger of arrest and death as a heretic on the one hand, and imprisonment as a lunatic on the other, the man of science in those days had a hard time. Columbus therefore sought far and wide for evidence to support his theories and render them acceptable. How much he learned — what, in the way of facts, he actually knew — it is hard to say. Having fallen in love with a Portuguese lady of good family, he married and apparently settled in Portugal as his home, but continued his voyaging. He knew the Mediterranean from end to end. He made several voyages to the Guinea coast, and dwelt for a time at El Mina, then newly founded, satisfying himself of the foolishness of the common assertion that men could not live "under the equinoctial" — that is, near the equator. He went north to and beyond Iceland, and acquainted himself with those waters, and thus convinced himself that the ocean was everywhere navigable, and subject to uniform laws of tides, weather, etc. His mind was cleared more and more of the mists of fable and superstition, and all he learned brought into clearer view the truth of science as a guide. He devoted more and more attention to improving the means of finding the true position of a vessel at sea, and of keeping a true course by the compass, which he continually studied; and it was he who first discovered that some leagues west of the Azores lay the meridian of no variation — a meridian that has now moved eastward until it lies near London. Everywhere he interrogated explorers, discussed navigation with experienced captains, and sought the aid of new maps, improved instruments, and advancing knowledge; and yet mixed with all seem to have been a childlike vanity, credulity, and superstition, hard to reconcile with his courage and acumen.

How much actual evidence he had of the existence of lands below the Atlantic horizon unknown to his countrymen can never probably be satisfactorily answered. The latest critical biographer of Columbus, the great Spanish liberal statesman Emilio Castelar, considers that he was led to his discoveries by little, if anything, outside of pure reasoning upon the rotundity of the earth and other scientific data, and dismisses as fables or things unknown to Columbus all the Scandinavian discoveries of Greenland and the rest, and other stories of men who, it is said, had already seen the transatlantic world he sought. We are told that he learned of woods and canes like none that grew in Africa, of strange carvings, and even of the dead bodies of men, resembling those of the far East, being cast upon the shores of Africa and the islands near it, especially the Azores. It seems impossible that when he was in Iceland and the other northern regions, a man of his inquiring mind should not have learned something of Greenland and the continental shores beyond, especially when one remembers that for centuries previous Catholic missionaries had been reporting progress to Rome from that distant but real field of labor. It is quite likely that some knowledge of these facts, which must have been known to the professors of the universities of Pavia and Barcelona, where Columbus studied, and to other intelligent men of Italy and Spain with whom he came in contact, had caused Columbus to go to the north, for we know of no other errand. Perhaps he had heard of the Zeni.

Especially to be noted is the allegation that Columbus possessed information as to the experience of a Frenchman named Jean Cousin, — a Dieppe sea-captain, who, it is asserted, discovered South America and the Amazon River in 1488. This claim has been lately reviewed ("Fortnightly Review," January, 1894)

by Captain Gambier of the British navy, and he decides that it is good; and that it was because Cousin's first mate was one of the Pinçons that that firm was willing to assist Columbus, as a good investment.

Whatever he knew or did not know, and whatever may have been the difficulties in his way, Columbus spent many weary years in fruitless efforts to interest some government in his schemes. How finally he won Spain to his support, secured the aid of the Pinçons, merchant princes of Palos, and sailed from that port on August 3, 1492,— and it was Friday! — are details that need not be repeated. Equally well remembered are the story of his daring onward voyage, and of the glorious outcome when, on October 12, land was seen,— a new world found.

Expedition after expedition followed one another from Spain to the newly found possessions, some conducted by the earlier companions of Columbus, and all filled with adventurers who cared for nothing but plunder. One of these, led by an officer named Ojeda, reached the coast of Guiana in 1499, and coasted along the north shore of South America as far, probably, as Maracaibo. This was the first of the Spanish expeditions actually to set foot upon the mainland; and it would not have been mentioned out of its place (since Cabot, as we shall presently see, had landed on the continent nearly two years before) but for the fact that one of its members was that Amerigo Vespucci whose fortune it was to have his name attached to the continent.

Amerigo Vespucci (or Vespusze, as Columbus spells it) was a Florentine engaged in the shipping business who was attracted to Spain by the maritime activity there, and became interested in equipping the second flotilla of Columbus and in other similar enterprises for the government. The wealth and influence thus gained and his general abilities led him to join that expedition of Ojeda in 1499, and during the next four years he made three other voyages to Brazil, in which the bay of Rio Janeiro was entered (New Year's day, 1501), and an exploration southward extended probably as far as South Georgia (Islands). Upon his return from this last voyage, in 1503, he publicly asserted that he had visited, in 1497, the coast of what is now the southern United States. It has lately been shown by Spanish records, however, that at that date he was busy in the government dockyards in Spain: therefore his assertion was false. It served, however, to deceive a forgetful public, and to procure for its author the coveted glory of being the first "discoverer" of the "New World," as he first called it (though there is no evidence that he understood it to be a continent), and hence the one entitled to give it his name.

This bold claim achieved its purpose. The oldest known map of the whole world, dated A. D. 1500, said to have been drawn by the great artist Leonardo da Vinci, from data furnished by Juan de la Cosa, and hence known to historians as the "De la Cosa Mappimundi" (it is preserved

in Madrid), bears the name "America" across the new countries for
the first known time; but Juan de la Cosa was with Ojeda and Vespucci on the expedition of 1499, and doubtless Vespucci managed the nam-
ing. In 1507, only a year after the death of Columbus, there appeared in
France the "Cosmographic Introductio" of Waldseemüller (also called
Hylacomylus), which was regarded as the most complete and authentic

THE "SANTA MARIA"—THE FLAGSHIP OF COLUMBUS' FLEET.

geography of its time; and here the name of *America* was boldly written
across "a fourth part of the world, since Amerigo found it." The name
(a Latin derivative) was novel, easy to pronounce, no one knew or cared as
to the right of it, and so it stood.

A few lines more as to the Spanish and Portuguese navigators in these
waters, and then we shall have done with them for the present. In 1499
one of the Pinçons sailed from Spain straight to the Amazon, as has
been mentioned, avoiding the West Indies, as if he knew precisely whither

he was bound, and reached there in January of 1500. A few months later a large Portuguese expedition under Pedro Cabral, starting for India around the cape, was blown so far to the west that it ran against Brazil. Everybody was hitting upon untrodden shores in those inspiring days, and Cabral promptly took possession for his king. As this shore was outside (east of) the hemisphere assigned by the Pope to the Spanish, the Portuguese kept it for 389 years, in spite of Pinçon's priority. In 1508 Ojeda obtained the government of the northern coast of South America, and Nicuesa of the region north of the Gulf of Darien; and with the arrival of these adventurers in New Spain began that era of rapine and horror which will forever disgrace the Spanish name. The rapacious governors and their wild crews quarreled and fought with each other as well as with the downtrodden natives, and exploration was carried on by piracy. A learned man, Martin Enciso, went out to take command in 1510, but he was deposed by his soldiers the next year and sent back to Europe, where he made the first book printed in Spanish (1519) describing America. His place was taken by Vasco Nuñez Balboa, who entered upon a career of exploration and peaceful conquest, generally conciliating the Indians, who told him of another sea not far to the west, and on September 25, 1513, guided him to the summit of a hill near Panama, whence he, first of Europeans, gazed upon the Pacific. Who can imagine the emotions of such a sight! — for it told the Spaniards that this land was not the eastern margin of Asia, but a new continent. Balboa made his way through the forests as rapidly as he could, and on the 29th, wading into the surf, banner in hand, took possession of the waters in the name of the King of Castile.

Balboa at once began preparations to utilize his discovery, for the Indians had also excited him and his men by tales of a country to the south abounding in gold. He cut and shaped timbers for small ships, and had with enormous trouble and labor transported these across the isthmus, intending there to build a fleet and sail southward, when he was superseded in command by a new governor, Pedrarias. This man, a jealous and brutal adventurer, on a false pretext of disloyalty arrested and beheaded Balboa before he could get away — an act that "was one of the greatest calamities that could have happened to South America at that time; for . . . a humane and judicious man would have been the conqueror of Peru, instead of the cruel and ignorant Pizarro." The frightful destruction of the country of the Incas soon followed, while Cortés overran Mexico and De Soto invaded Florida.

It has doubtless by this time been in the mind of more than one reader to ask whether, while the men of the Mediterranean region were making

A PEACEFUL DAY ON THE SPANISH MAIN.

these notable searchings for new shores, the men of northern Europe were standing idle. What were the mariners of France and the Netherlands, Scandinavia and Great Britain, doing? Well, all were doing something, and some of them produced results of novel seafaring that were well worth the getting, but these were principally in far northern waters, as we shall read in the next chapter. It was not until the opening of the sixteenth century that in England, at least, that era of far voyaging began which signalized the Elizabethan age on the sea as much as the poets and dramatists and statesmen-writers of her court distinguished it on land.

It was, however, earlier than that—in the reign of Henry VII—that England's story of discovery begins, and the first names are those of two Italians known in English as John and Sebastian Cabot, father and son, who were then residents of Bristol. The Bristol folk were at that time the foremost mariners of England, who often went to Iceland and all the nearer isles: and they firmly believed in certain traditional islands and coasts far away to the west, which seem to have been composed of no better material than the airy structures of the sunset clouds and the romantic tales of Phenician sailors and other travelers in the dawn of history. As long ago as when Strabo wrote, a century before the birth of Christ, these things were of old belief, and he recounts the delights then told of the "Isles of the

Blest," west of the farthest verge of Africa. When the Canary Islands became known as facts, the myth moved farther west; and when acquaintance
with the Azores proved them to be only natural earth with a fair share of
its ills, as well as of its good, people insisted that still other islands must lie
farther away, where the Elysian Fields basked in perpetual summer and men
were eternally happy. The old idea charms us even yet when we sing
"Sweet fields beyond the swelling flood stand dressed in living green."

But no such higher rendering occurred to the men of the earlier time.
They believed firmly in the actual existence of these ever-fortunate islands
under the sunset horizon of the Atlantic, and (in the north) called them the

VOYAGING TO THE ISLES OF THE BLEST.

Isles of St. Brandon, the "green isle of Brazil" (the root of which word
seems to express the idea of redness, such as appears in low sunset clouds),
the Isle of the Seven Cities or Antillia, and by other names. Ferdinand
Columbus, a son of Christopher, says in his "History" that his father fully
expected to meet, "before he came to India, a very convenient island or continent from which he might pursue with more advantage his main design."
This does not prove that Columbus put any faith in the reality of these old
notions, nor does he seem to be responsible for the fact that the name *Antilles* was immediately attached to the archipelago he actually did meet with,
and *The Brazils* to a part of the mainland next found. These names had
been appearing on conjectural maps of the Atlantic side of the earth for

many years before his time; and that they represented realities to many hard-headed merchants and sailors of his time is shown conclusively by the fact that between 1480 and 1487 at least two carefully planned naval expeditions had gone from Bristol, England, in search of them. How much vague memories of early Norse and Irish findings in the west may have given weight in Bristol to these old myths is hard to say; but at any rate it was there the search for this pot of gold at the end of the rainbow bore unexpected and momentous results, but all were surprised at the distance involved.

About 1496 John Cabot, then a resident of Bristol, proposed to the king an expedition in search of a new route to the Indies by sailing due west from Ireland. Henry VII was excited by the news of Columbus' southerly findings, and was eager to secure something of the kind for England. Nevertheless, although the king granted privileges that might prove profitable in case of success, he seems to have furnished no money. Cabot, therefore, sailed away, privately equipped, in a small caravel named *Matthew*, carrying only eighteen persons.

Never was a voyage of discovery, the consequences of which were so far-reaching, entered upon with less pomp or flourish of trumpets. So little note of it was made at the time that the very name of John Cabot narrowly escaped being lost altogether, and the record of his work came very near being replaced by a confused account of the doings of his son Sebastian; for it was not until certain letters had been found—and that within a very few years—in the contemporary archives of Spain and other European countries, that we were able to give any sure account of the matter.

It is now plain that John Cabot, in the *Matthew*, leaving Bristol early in May, 1497, and having passed Ireland, shaped his course toward the north, then turned to the west and proceeded for many days until he came to land, where he disembarked on June 24, and planted an English flag.

There seems to be no doubt that this was the mainland of North America, and the general opinion has prevailed that his landfall was the extremity of Cape Breton. Cabot stayed some days, but how far he traced the coast, and whether he learned of Newfoundland or Prince Edward Island, are matters of conjecture. At any rate, he soon turned homeward, and arrived in Bristol probably on August 6.

We can imagine with what eagerness his story was listened to, as he told of the fair, temperate, well-wooded land, its people and animals and fruitfulness, that he had seen. But the thing that impressed the Bristol men most was the report of the enormous abundance of codfish there. This was something these canny men could see without any illusions, and possess themselves of regardless of papal bulls; and they at once aban-

doned their northern fishing-grounds and began to resort to the Banks of
Newfoundland, whither they were quickly followed by large annual fleets
of Norman, Breton, Spanish, and Portuguese fishermen. John Cabot
intended to go again the next year and make his way onward to Japan, as
he believed he could do, for, like the others, he thought what he had found
was only a remote eastern part of Asia ; and in 1498 he actually did sail westward from Bristol with five ships, victualed for a year. None of these ships
ever returned, and no evidence exists that they ever reached their goal ;
and with them John Cabot, to whom England owed her early supremacy in North America, disappears from view.

Sebastian Cabot was a son of John Cabot, and a skilful map-maker.
Whether he went with his father on the first voyage is disputed ; there
seems no direct evidence that he did so. That he did not go on the second
voyage is plain, for he had a long subsequent career, of which accurate
knowledge is a late acquisition ; here it is only necessary to add that by his
statements to Peter Martyr and others he allowed the erroneous impression to pass into history, if he did not directly authorize the lie, that it was
he, and not his father, who discovered America and the fishing-grounds.

Now that the way across the Atlantic was learned, chivalrous sailors
hurried to add what they could to the map. Corte-Real, a Portuguese
of rank, struck northwest, and hit upon and named Labrador as early
as 1500. The next voyage of prominence introduces the French as competitors, Francis I sending the Florentine Verrazano, a typical sea-rover
of the period, who had already been to Brazil and the East Indies and
was finally hanged as a pirate, to find out what he could about northern
America. He steered west from the Madeira Islands in January, 1524,
found land near Cape Fear (North Carolina), and claimed to have traced
the coast as far north as Nova Scotia, besides entering a large bay (either
New York or Narragansett). His whole story, however, rests on certain
letters and maps the authenticity of which has been hotly disputed ; and
at any rate little, if anything, came of this voyage.

It was far different with the next one, however,— that one sent from
France in 1534, under the command of Jacques Cartier, who sailed from
St. Malo in two tiny vessels to Newfoundland, and learned of the mouth
of the St. Lawrence River. Then, like all the other captains, none of
whom could stay over winter in America, because their vessels were too
small to store provisions, and because they were beset by fears, not only
of visible savages, but of invisible hobgoblins, he returned to France. The
next year found him back again, however, this time steering his vessels up
the St. Lawrence to "Hochelaga" (Montreal), and later carrying home an

account that led to so immediate a movement on the part of France that Canada was the scene of the earliest colonization of the New World, properly speaking, for the Spanish settlements in the south were thus far nothing but military stations. France, indeed, dreamed of obtaining the whole of North America for herself, and attempted soon after to colonize Florida and the Carolinas; but these attempts failed, and she was able to hold only the valley of the St. Lawrence and the shore of its gulf. These things happened later, however, and for many years the Atlantic coast of North America was left unclaimed by any one, while the English and Dutch were busy in the far north, the Spanish were rioting in the tropics, and the Portuguese turned their attention to the southern and eastern quarters of the globe. It is one of the most striking curiosities of the history of the development of civilization on the globe, following the stagnation of the middle ages and the desolation of the plague-ridden thirteenth century, that the most remote, unprofitable, unhealthy regions were so fiercely struggled for, while the best parts of the New World were left until the last.

Having found Brazil, both Spaniards and Portuguese proceeded to trace the continent southward, hoping to find a practicable way to Peru around it. Several experienced navigators worked southward, the best known of whom is Juan Diaz de Solis, who entered the La Plata River and was killed there by the Indians in 1516. Columbus had not been a moment too soon to be first. Nevertheless it was left to a stranger in those waters, the indomitable Magellan (Fernão de Magalhães), to reap the reward of success. The Pope and all the bishops still declared that the earth was flat; but so little was this now believed, even by themselves, that Magellan, who had just quitted the service of Portugal, dared to propose to "his most Catholic majesty" the King of Spain to sail west instead of east to the Moluccas, just as though the earth were globular and might be circumnavigated; and the king not only dared to listen, but approved of the proposition, which seemed entirely practicable *if* South America could be passed. That was the problem Magellan set himself to solve. Should he succeed, could the Moluccas be reached by sailing westward, then they would fall into that half of the earth given by the Pope to Spain, and Portugal's present claim to them would be overthrown. Thus the experiment was well worth making in behalf of politics as well as knowledge, and Magellan was furnished with five ships, carrying two hundred and thirty-seven men. The *Trinidad* was the admiral's ship; but the *San Vittoria* was destined for immortality.

He struck boldly for the Southwest, not crossing the trough of the Atlantic as Columbus had done, but passing down the length of it, his aim being to find some cleft or passage in the American continent through which he might sail into the Great South Sea. For seventy days he was becalmed under the line. He then lost sight of the North Star, but courageously held on toward the "pole antarktike." He nearly foundered in a storm, "which did not abate till the three fires called St. Helen, St. Nicholas, and St. Clare appeared playing in the rigging of the ships. In a new land, to which he gave the name Patagoni, he found giants of "good corporature." . . . His perseverance and resolution were at last rewarded by the discovery of the Strait named by him San Vittoria, in affectionate honor of his ship; but which, with a worthy sentiment, other sailors soon changed to the *Straits of Magellan.* On November 25, 1520, after a year and a quarter of struggling, he issued forth from its western portals, and entered the Great South Sea, shedding tears of joy, as Pigafetta, an eye-witness, relates, when he recognized its infinite expanse. . . . Admiring its illimitable but placid surface, and exulting in the meditation of its secret perils soon to be tried, he courteously imposed upon it the name it is forever to bear — the *Pacific Ocean.* . . .

And now the great sailor, having burst through the barrier of the American continent, steered for the northwest, attempting to regain the equator. For three months and twenty days he sailed on the Pacific, and never saw inhabited land. He was compelled by famine to strip off the pieces of skin and leather wherewith his rigging was here and there bound, to soak them in the sea, and then soften them with warm water, so as to make a wretched food; to eat the sweepings of the ship and other loathsome matter; to drink water that had become putrid by keeping; and yet he resolutely held on his course, though his men were dying daily. As is quaintly observed, "their gums grew over their teeth, and so they could not eat." [This was scurvy, the dread of all mariners in those times and long afterward.] He estimated that he sailed over this unfathomable sea not less than 12,000 miles.

In the whole history of human undertakings [declares Dr. John W. Draper, from whose striking sketch of this achievement in his "Intellectual Development of Europe" I am quoting] — in the whole history of human undertakings there is nothing that exceeds, if, indeed, there is anything that equals, this voyage of Magellan's. That of Columbus dwindles away in comparison. It is a display of superhuman courage, superhuman perseverance — a display of resolution not to be diverted from its purpose by any motive or any suffering. . . .

This unparalleled resolution met its reward at last. Magellan reached a group of islands north of the equator — the Ladrones. In a few days more he became aware that his labors had been successful. He met with adventurers from Sumatra. But, though he had thus grandly accomplished his object, it was not given to him to complete the circumnavigation of the globe. At an island called Zebu, or Mutan, he was killed, either, as has been variously related, in a mutiny of his men, or, as they declared, in a conflict with the savages, or insidiously by poison. . . . Hardly was he gone when his crew learned that they were actually in the vicinity of the Moluccas [having previously wandered too far north, and discovered the Philippines], and that the object of their voyage was accomplished. . . .

And now they prepared to bring the news of their success back to Spain. Magellan's lieutenant, Sebastian d'Elanco, directed his course for the Cape of Good Hope, again encountering the most fearful hardships. Out of his slender crew he lost 21 men. He doubled the Cape at last; and on September 7, 1522, in the port of St. Lucar, near Seville, under his orders, the good ship *Vittoria* came safely to an anchor. She had accomplished the greatest achievement in the history of the human race. She had circumnavigated the earth.

The immediate result of this voyage was to impress Spain's sovereignty upon the East Indies; but a vaster and more far-reaching consequence was the influence it exerted, by its proof that the world was really a globe, to free men's minds from blind belief in and guidance by a tradition, which had taught that the earth was a flat plain surrounded by water,—an error sanctioned by St. Augustine and other influential teachers. Magellan impressed a name upon the greatest of the oceans, and has his own name gloriously emblazoned upon both the map of the earth and the map of the sky in the southern hemisphere; but his greatest title to honor, after all, is that he struck dogma the hardest blow it ever received.

The sixteenth century seems to have been, outside of the Arctic regions, an era of surveying rather than of exploration by sea, yet some notable

SEA-SURF AT SANTO DOMINGO.

work was done in the East, where all nations now entered as competitors in the trade, seizing upon every island or mainland shore that they could, and holding their possessions as long as possible. Even the English entered heartily into this rivalry, the great East India Company having been founded in 1599. With its trading we have nothing to do, but must note that it extended knowledge of Oceanica considerably, and added greatly to Europe's information as to India, the Malayan peninsula and larger islands, China, and Japan. The Spanish and Portuguese found themselves so busy in defending that to which they already laid claim that they had little time to search for new lands; and this sort of enterprise fell mainly to the Dutch, who, now that the Netherlands were at last free from the long and cruel tyranny of Spain, were energetically making up for lost time. Their captain, Van Noort, went out by way of the Straits of Magellan to the Philippines, and got back to Rotterdam between 1598 and 1601. Another fleet made the same voyage fifteen years later; and in 1616 Cape Horn was rounded by Willem Cornelis Schouten, who gave the name of his home village, Hoorn, to that desolate terminus of South America.

For many years geographers had held belief in a vast "southern continent,"— *Terra Australis*,— and most of the islands found in the South Pacific were accidental results of some attempt to reach it. New Guinea had been sighted a century before, and perhaps Australia also, of which several navigators got glimpses here and there early in the sixteenth century, satisfying them that it also was a great island. It was not until this century was half gone, however, that the map of that quarter of the "South Sea" was filled out with any accuracy; and this was due to the skill and labor of an eminent Dutch voyager, Abel Janszen Tasman, who was despatched southward with two ships by the colonial government at Batavia, where the Dutch had already gained political ascendancy.

"This voyage," we are told, "proved to be the most important to geography that had been undertaken since the first circumnavigation of the globe." Tasman sailed from Batavia in the yacht *Heemskirk*, on the 14th of August, 1642, and from Mauritius on the 8th of October. On November 24 high land was sighted in 42° 30′ S., which was named Van Diemen's Land (now Tasmania), and, after landing there, sail was again made, and New Zealand (at first called Staatenland) was discovered on the 13th of December. Tasman communicated with the natives and anchored in what he called Murderers' Bay, because several men were massacred there by the natives. Thence he took an irregular course east and north, until he arrived at the Friendly Isles of Cook. In April, 1643, he was off the north coast of New Guinea, having meanwhile sailed around New Britain and New Ireland (now New Mecklenburg), and on June 15 he returned to Batavia.

The contribution to sea-knowledge of the remaining voyages in this century were mainly in the direction of a better understanding of winds, currents, ice movements, tides, and an improvement in the methods of building, rigging, and navigating vessels intended for long voyages. Map-making received a great impetus and was especially cultivated by the Dutch, among whom Mercator became famous by inventing the useful projection that bears his name and is still most commonly used. Nevertheless, the improvement, especially in instruments of navigation, was slow,

The astrolabe generally gave place to the cross-staff; and this to a better device called the back-staff, of which an improved form, invented by John Davis, remained long in use. This was called the Davis quadrant; and with it "the observer stood with his back to the sun, and, looking through the sights, brought the shadow of a pin into coincidence with the horizon." Many variations of this instrument were made, until, in the middle of the next century, it was superseded by the sextant. Thus before the close of the seventeenth century, astronomers and navigators had learned how to determine latitude fairly accurately, and the sailor had prepared for him a variety of tables of stars, almanacs, and other mathematical guides. The determination of longitude was yet difficult, however, owing largely to the imperfection of timepieces; and it was not until the last year of the century, signalized by the first recorded sea-voyage made purely for scientific purposes, that much advance was made. This voyage, lasting two years (1699–1700), was undertaken by the eminent English astronomer Edmund Halley for the express purpose of obtaining information necessary to the improvement of the compass and methods of ascertaining the position of a ship at sea, was productive of results of the greatest service, and placed the science of navigation upon a sure footing. It was followed early in the next century by the establishment in England of the Longitude Board, a scientific commission charged with the duty of determining longitudes and studying navigation. From this board came the "Nautical Almanac," which first appeared in 1767, but similar almanacs are now published annually by the governments of almost all maritime powers, and the editorship is esteemed in the United States one of the most honored positions in the naval service. These books contain ephemerides, or tables of positions for each day of that year of all the heavenly bodies, "predictions of astronomical phenomena, and the angular distances of the moon from the sun, planets, and fixed stars," all referred to some stated meridian.

With such an almanac, an improved compass, and one of Newton's new sextants as a means of quick and accurate observation of sun and moon and stars, the navigator had little need to doubt as to where he was; and maps began to show a corresponding improvement in accuracy.

The early part of this century, as we shall see later, was the era of the buccaneers and of many wild sea-rovers whose far-wandering barks, in search of adventure, picked up much information at the expense of human lives and hard-earned property. The foremost of these was Dampier, who seems to have gone almost everywhere a ship could go, and who found out many new things, which he had the power of telling well, as to Australasia; and the strait between New Guinea and New Britain, which he

discovered, is named after him. Many a commander was now cruising in
those waters, however, under English and Dutch flags mainly, finding new
lands and pillaging old ones — such as Roggewein, Anson, Byron, Wallis.

"BUCCANEERS AND MANY WILD SEA-ROVERS."

Carteret, Bougainville, and others; and such important islands as Easter,
Tahiti, Charlotte, and Gloucester groups, Pitcairn, and others, were found
during the first half of this century. But now the English were to redeem
their good name by sending out a government expedition, or series of expeditions, whose object was scientific discovery and the humane study of the

men and resources on the other side of the world, instead of forced trade or bloody rapine. These were the three expeditions commanded by Captain James Cook, one of the most capable officers in the British navy.

The first voyage was made in 1767 for the purpose of carrying a party of astronomers and naturalists to Tahiti to observe there a transit of Venus, after which a survey was made of the then almost unknown coasts of New Zealand and Eastern Australia. The second voyage was to explore the Antarctic regions, whither the French had preceded him, as we shall see in the next chapter; and we need only say here that Cook finally disposed of the tradition of a vast *terra australis* — at any rate a habitable one. It is to his third voyage, then, that he principally owes his fame.

This was undertaken in pursuance of the ruling idea of his day, that a sea-route might be discovered north of the American continent, which would vastly shorten the trip from Europe to China and the Spice Islands. Others were seeking it directly by way of Baffin's Bay, and Cook was sent to attack the problem from the Pacific side. He was given command of his old ship *Resolution* and a new one, *Discovery*, outfitted in the best possible manner, and especially guarded, in the matter of provisions, against scurvy — that dread of the old-fashioned seamen, in respect to which Cook himself had introduced such new and valuable preventives as would alone have entitled his name to grateful remembrance. He was commanded to revisit, on his way out, the South Pacific Islands; and departed from England in June, 1776, steering by way of the Cape of Good Hope, and reaching the archipelagoes in the spring of 1777, where he cruised for nearly a year. In January, 1778, he sailed north from the Friendly Islands, and a few days later hit upon a large, inhabited, unknown group of islands, whose principal one was called Hawaii by the people, but which he named Sandwich Islands, in honor of an English earl who had taken great interest in his plans. Here he spent some delightful days, and then bore away to the west coast of America, which the English still claimed under Drake's name of New Albion, and which he struck near Puget Sound. Thence he went slowly along the coast northward until he found and penetrated the deep bay since called Cook's Inlet. His hope that this might prove a sort of northern Straits of Magellan was quickly disappointed, and he went on into and through Bering Sea and Strait until he was stopped by the ice on the north shore of Alaska, at a point still called Icy Cape. Then he turned back, surveying both shores of the strait, and again made his way to the Sandwich Islands, where, in an unfortunate quarrel with the natives, he was killed. The Hawaiians have always said that this was the act of a ruffian among them, and that the chiefs and the best of the people never wished nor intended any harm to their visitors; and this is probably true. His executive officer took the ships back to England in October, 1780.

The voyages of this able and intelligent commander bore fruit in many ways. One was the colonization of Australia, Tasmania, and New Zealand by the English, which began in 1788. Another was the voyage of Vancouver to the west coast of America in 1792, which intercepted the surveys of the Spaniards there, under Quadra and others, and enforced English possession of all the country between the Californian settlements of the Spaniards and the Russian posts in Alaska, though he curiously failed to find either Puget Sound or the Columbia River. A third direct result, and from some points of view the most important one, was the opening of a great number of the South Sea Islands to Christian missionaries.

By this time there had arisen in the New World which these voyagers had first stumbled upon, and then searched for, and afterward scrutinized so carefully, a new, composite nation, which somehow forgot that all their broad and fertile land had been given away centuries before by an old gentleman in Rome to his friends the Spaniards, and acted as though they thought it belonged to themselves; and by and by this thriving nation hoisted a starry flag of its own, and proclaimed itself the United States of America. Then, not to be behind European powers, whose navigators were enriching libraries with magnificent chronicles of scientific studies in sea-science, such as those of the French "Voyage of the *Astrolabe*," the Russian narratives of Krusenstern and Kotzebue, and the English explorations of Beechey (who was accompanied by Charles Darwin), the United States sent to the Pacific a well-equipped expedition under Lieutenant (afterward Admiral) Charles Wilkes. This was gone from 1838 to 1845, surveyed the west coast of South America, wandered about Oceanica, and did its best to penetrate the icy limits of the Antarctic zone. The results were six magnificent folio volumes, containing not only the narrative of the cruise, but contributions to science by James D. Dana, Horatio Hale, John Cassin, and other men of the last generation great in American science.

CHAPTER V

SECRETS WON FROM THE FROZEN NORTH

S soon as the sea-routes between Europe and the far East were learned, and the American coasts had been mapped, the region within the Arctic circle became the most attractive field for nautical discovery. All this earlier Arctic exploration, however, was not, as it has lately become, a system of scientific research, but was simply a series of attempts to open new roads for commerce to follow. It occurred to every navigator that as a sea-way had been gained past the southern end of America, so one around its northern border might be disclosed; and perhaps, also, a ship-route along the northern coast of Siberia. Either of these would be far shorter than to go to "Cathay" around either Cape Horn or Cape of Good Hope, and would enable the English and other northerners to avoid their enemies, the Spanish and Portuguese, who commanded the southern waters.

The first Arctic voyage of exploration, properly speaking, was that of Willoughby and Chancellor, who in 1553 penetrated the seas north of Scandinavia, where they became separated. Willoughby and his men tried to winter on the coast of Russian Lapland, but all died of scurvy.

Chancellor, however, pushed on into the White Sea, reached a monastery on the coast, and thence made his way to Moscow, where he was well received, and thus opened a trade route of incalculable advantage to both England and Russia. It led at once to the organization of the Muscovy Company, and began a commerce now regularly carried on in steam vessels to Archangel, which in 1897 was connected with Moscow by railroad.

By 1580 several other commanders had tried to improve on this performance, but none got past the Kara Sea, and the next important effort was headed toward that "Northwest Passage," which for more than three centuries was the lodestone of Arctic students and voyagers. It was in charge of Martin Frobisher, later one of England's most conspicuous admirals, who afterward made a larger expedition in which he learned many facts about the Labrador coast and Hudson Strait. Another English seaman, and a more scientific one, John Davis, made three remarkable voyages, between 1585 and 1589, and increased the map by a careful delineation of both coasts of the strait still called after him.

Shortly afterward Dutch merchants had sent three expeditions northward under command of William Barentz to search for a *northeast* passage, the third and most important of which sailed in 1596 and found it impossible to penetrate the ice east of Nova Zembla (which had been seen first by Burrough in 1556, who had been shown the way by Russian fishermen), but discovered Bear Island and Spitzbergen. The crew of Barentz's vessel spent the winter of 1596–97 at Ice Haven, Nova Zembla—the first successfully to face a winter in the Arctic zone. When the next spring came they made their way to Lapland and homeward in boats, but Barentz died on the road. This voyage was highly important in opening to the Netherlands the whale and seal fisheries of that region which has ever since been known as Barentz's Sea, but it discouraged the hopes of a "northeast passage." In 1871 Barentz's winter quarters at Ice Haven were found undisturbed, after a lapse of 274 years, and in 1875 part of the journal kept by this brave mariner was recovered. Almost every year about this time saw English, Dutch, and Danish ships going north, each adding some new fact to geography and the knowledge of polar waters and ice. One of them, in 1607, was commanded by Henry Hudson, who searched the North Atlantic, found Jan Mayen, and pointed the way to the Spitzbergen whale fisheries; yet he had hardly more than a sail-boat, and a crew of only eleven men.

The following year this intrepid man tried to go to China north of Asia, but failed as Barentz had done, and returned "void of hope of a northeast passage." Nevertheless, he tried it again a year later in the service of Amsterdam merchants, but his men were obstreperous, and, yielding to his own inclination as well as to theirs, he turned west to find that "Northwest Passage" in which everybody then believed because they hoped, and because of the difficulty of getting so great a fact as the real North American continent proved to be accepted by the popular imagination, which was used to small things in geography. Very willingly, then, Hudson's little ship, the *Half Moon*, was turned toward the southwest; and it found something better than it sought, for the Hudson River and the site of the future metropolis of the New World were added to the map.

Hudson's success in this voyage led to his immediate engagement by a company of English merchants and speculators, who were willing to risk additional money in searching for a northwest passage if he would lead.

In 1610, therefore, Hudson took command of a new ship, the *Discoverie*, and sailing in her to Baffin's Bay, found the great opening of Hudson Strait, and with high hope that his goal was now in view followed it westward into Hudson Bay. Here he coasted south to what we term James Bay, and, after a comfortable winter, resumed his examination of the west coast, whereupon the majority of his men mutinied, set Hudson and several sick men adrift in a rowboat, and turned back. Most of the mutineers died, but the vessel was finally taken back to London, where the murderers were promptly questioned and nearly as promptly hanged.

The story of another remarkable voyage closes the story of this early attempt at the problem which, two hundred and fifty years afterward, was to be solved only by proof of its uselessness. In 1616 another *Discovery* — a caravel of only fifty-five tons—went north from England in charge of William Baffin. "On the 30th of May he had reached Davis' farthest point, Sanderson's Hope, in 72° 41′ N., ... and reached, 1st July, an open sea, the 'North Water' of the whalers of to-day. Passing Capes York, Atholl and Parry, he yet pushed northward, and on 5th July attained his farthest point within sight of Cape Alexander. His latitude, about 77° 45′ N., remained unequaled in that sea for two hundred and thirty-six years." Arctic success depends on good luck.

The next century (1700 to 1800) was a period of active polar research in the Old World. The Russians completed their knowledge of their Arctic coasts, Popoff reaching East Cape in 1711, and bringing back an account not only of various islands, but also of a continental shore eastward.

FANTASTIC ICEBERGS IN HUDSON STRAIT.

It was this report that caused Peter the Great to set on foot a costly scheme of research upon the northeastern coasts of Siberia, which was placed in the hands of Vitus Bering, a Dane, in his navy, but accomplished nothing of any value; and it was not until 1740 that Bering finally crossed over in a blundering sort of way and made a brief examination of the coast of Alaska, where his ship was finally wrecked, and he died of discouragement and chagrin. He saw neither the sea nor the strait that bears his name, was not the first to reach the American continent, and never learned whether or not it was connected with Siberia. Nevertheless his voyage had fruitful results, for it led to vast fisheries and fur-gatherings, and the writings of his naturalist, Steller, had and still have great scientific importance.

A WALRUS BREEDING-GROUND, BERING STRAIT.

By this time the whaling and allied marine industries, and the work of such excellent explorers as the Dutchman Martens, had made mariners thoroughly acquainted with the North Atlantic from Nova Zembla to Greenland, and a vast advance had been effected in the knowledge of navigation amid the ice, and in the building and equipment of ships and the proper methods of provisioning and clothing and treating crews in order to maintain health and comfort as well as mere safety. These well-fitted and daringly managed whalers had at the beginning of the nineteenth century begun to penetrate far into the waters west of Greenland, in spite of a very curious fact, which would make anybody but a British whaleman pause — namely, that there were no such waters. So their best maps and treatises said!

Two hundred years had now passed since Baffin's return from his wonderful voyage of 1616, and during all that time not a white man's keel had plowed the chilling solitudes he had left, except lately these venturesome whalers, who did not frequent libraries. Consequently Baffin's work had first been forgotten and then disbelieved; so that at last first-class maps were published which omitted Baffin's Bay altogether, and books were written, such as Barrows' "Arctic Voyages" (London, 1818), that denied the authenticity of his narrative. As the nineteenth century opened, however, England began to turn her attention to the renewal of polar studies. The Hudson's Bay Company's men were reaching the coast of their Territories here and there; but otherwise the whole Arctic Ocean north of British America was unknown.

To relieve herself of the shame of this Great Britain soon sent into that field a rapid succession of explorers, many of whom soon became famous. The very first of these, John Ross, despatched in 1818, confirmed fully the geography laid down by Baffin as far as Cape York, in spite of the learned book-makers, and reported a great number and variety of interesting facts; whereupon a much larger expedition was at once arranged and placed in command of a naval officer named William Edward Parry, who went out in 1819 with orders to find the northwest passage, and who had in his staff such men as Sabine, Liddon, James Ross, Reid, Crozier, and similar material, all stimulated not only by naval and scientific pride, but by the offer by Parliament of a reward of $100,000 to him who should first discover the desired thoroughfare.

This first voyage was a grand success. Forcing his way into Lancaster Sound in midsummer, Parry found that Ross's report that it was a land-locked bay was erroneous. As Greely tells it :

> The mirage-mountains of the previous year had vanished, and as Parry crowded sail westward, he opened a series of magnificent waterways hitherto unknown. The way lay through an archipelago (Parry), with North Devon, Cornwallis, Bathurst and Melville islands to the north, and Cockburn, Prince of Wales, and Banks islands to the south. Lancaster Sound, broken at its western end by Prince Regent Inlet, gave way to Barrow Strait, which broadened into Melville Sound, while yet farther to the west the encroaching land formed Banks Strait wherethrough these channels open into polar ocean.

If you will look at the map you will see that this list comprehends pretty nearly everything south of Smith Sound. Many details of course were lacking, and these Parry was sent a second time to work out, but he added really little to geography by two seasons of hard work; and a third voyage, begun in May, 1824, was still more unfortunate. These voyages, however, enabled Parry, who was one of the greatest of all Arctic students and navigators, to state that the western sides of all northerly and southerly bodies of water are always more encumbered with ice than the eastern sides; and to make many most valuable improvements in ice navigation and equipment. His illustrated narratives remain among the most readable books of Arctic experience, and little has been added to their accounts of eastern Eskimo life and customs.

Meanwhile (1819) another navy officer, who was ardent in the scientific branches of his profession, as well as distinguished in seamanship and naval warfare, and who had acquired Arctic experience under Buchan in the ill-starred expedition of 1818, was sent overland to coöperate with others in defining the mainland coast of America. This was Lieutenant John Franklin — a name destined to become the most famous of all among the explorers of the frozen North. For several years he and his parties lived and traveled among the Eskimos, tracing the coast-line from a considerable distance east of the mouth of the Coppermine River westward almost to Point Barrow, Alaska, where they came within one hundred and forty-six miles of meeting Beechey's coöperative examination by sea from Bering's Strait; and it was out of these trips that we got the valuable treatises upon the natural history of British America, published by his assistants, Hearne and Richardson. This ended in 1826.

The next prominent expedition was that of Captain John Ross and his nephew James, afterward celebrated in Antarctic exploration; and it turned out an exceedingly productive one. Meeting fortunate conditions in Lancaster Sound he easily reached where the *Fury* had gone ashore, and refilled his ship with a portion of the stores Parry had thoughtfully landed and made safe there — a provision which later kept this expedition from destruction. Then he pressed on beyond where Parry had gone, and added largely to the details of his map, but curiously failed to recognize Bellot Strait as a thoroughfare, and so unaccountably missed the thing he was in search of. Ross discovered Boothia Felix; and during the three winters spent on its eastern shore, the younger Ross, by sledging, discovered Franklin Passage, Victoria Strait, and King William's Land, and largely explored their coasts; but his most important work, "giving imperishable renown to his name," as Greely declares, was the determination of the position of the north magnetic pole on the west coast of Boothia Felix.

"The experiences, duration, and results of this voyage," writes General A. W. Greely, "are among the most extraordinary on record. The party passed five years in the Arctic regions without fatality, save three (two from non-Arctic causes), discovered a new land, the northern extremity of the continent of America, and made other extensive geographical discoveries. Its observations are probably the most valuable single set ever made within the Arctic circle."

ESKIMOS IN SUMMER
TENTS.

During the third winter (1833) a rescuing party under Captain C. Back
had gone from England overland in search of Ross; and recruited by Hudson's Bay Company men of experience had descended Fish (or Back's)
River to its mouth, thus noting a new point on the map; but it failed to
reach Ross. By similar overland journeys from their trading-posts on
Great Slave Lake and elsewhere, the Hudson's Bay Company's men, especially Simpson, Dease, and Rae, connected various points of the coast,
so that before 1850 it was known with substantial accuracy from Melville
Peninsula to Bering Strait. In much the same way Russian sledge-travel-
ers had traced the northern Asiatic coast by descending to the mouths of
rivers; but no ship had yet succeeded in passing Cape Chelyuskin, the
northernmost point of Asia or any continental land.

Then came a period of the keenest rivalry and richest results in the
history of polar conquest, but also one of the greatest catastrophes. The
expeditions of Lieutenant John Franklin in 1818 and 1819 were spoken of a
moment ago. His services then and subsequently had been recognized by
the British king, who, among other honors, had made Franklin a knight,
and sent him to be governor of Van Diemen's Land (Tasmania), where he
remained from 1836 to 1843, founded a prosperous colony, and was regarded
as one of the wisest, kindest, and most upright men of his day. Upon his
return to England Franklin was made commander of the most important
expedition that had ever yet been fitted out to search for the Northwest
Passage, and his reputation brought the best men as volunteers to his standard. Having selected 134 officers and men, and made the best equipment
possible, Captain Sir John Franklin sailed on May 19, 1845, in the *Erebus*
and *Terror*, Parry's old ships. On the 26th of July they were seen pro-

ceeding prosperously up Baffin's Bay by a whaler, who reported them in due course, but neither ships or crews were heard of again for many years.

Anxiety over the long silence at length aroused the people of England and the United States to begin a search for them which lasted through many years. It was fruitless as to its first object,— the rescue of Franklin or any survivors,— but it gradually cleared up the sad mystery, and it was the means of learning all, and more than all, that Franklin sought to ascertain.

The search began by the despatch, early in 1848, of Sir James Ross in two ships, *Investigator* and *Enterprise*, which wintered near the northeast point of North Devon, and returned the following year with no tidings, although they afforded the second officer, Lieutenant F. R. M'Clintock, an opportunity to acquire a knowledge of sledging, which he afterward used to advantage. This failure only aroused England to renewed efforts.

Many ships were started out at once, and also parties overland, of which mention will be made later. The *Herald* and *Plover*, during 1848 and 1849, scanned the whole coast from Bering Sea to the mouth of the Mackenzie, and discovered Herald Island. Following them, in March, 1850, went the *Enterprise*, under Collinson, and the *Investigator*, under M'Clure, via Bering Strait, while the *Assistance* and *Resolute*, with two steam tenders, under Captain Austin, went to renew the search by Barrow Strait, and two brigs, the *Lady Franklin* and *Sophia*, under a whaling captain named Penny, followed them. The eastern expeditions discovered Franklin's winter quarters of 1845–46 at Beechey Island, but no record of any kind indicating the direction taken by his ships. Admirable arrangements were made for passing the winter, and their combined sailing and sledging work added much to the map of that district, and to our knowledge of life in polar latitudes, but it learned nothing whatever of Franklin's fate.

Meanwhile the expedition via Bering Sea had become separated in the Pacific, and M'Clure, in the *Investigator*, got so far ahead that he was able to pass through Bering Strait and work his way eastward north of British America, and through the narrow Prince of Wales Strait until he reached Princess Royal Islands, where he wintered. Here he was only thirty miles from Barrow Strait; and when he had climbed a high hill and saw its ice gleaming in the distance, he had in reality discovered the Northwest Passage. Yet he was not the first, as we now know, for when the survivors of Franklin's ships, in their attempt to escape, had reached Cape Herschel, they, too, saw this same passage they had been sent to find, but then, as now, it was closed by perpetual ice, so that although we now know the way, we can no more avail ourselves of it than could they, except by going south of King William's Land, through a strait of which they had

A FLOATING ICE-CASTLE OF THE FROZEN NORTH.

"Out from the dark, mysterious North,   Tingling with unforgotten dreams,
With all its glamour, every night   And every day flood-full of light."

not yet learned. The next summer was spent in a fruitless struggle to get north along the western side of Banks (or Baring) Land, in which he succeeded only far enough to get frozen in so firmly on the north shore of that great island that even the summer warmth did not release his ship. . He would have perished had it not been that musk-oxen were plentiful; and by the spring of 1853, it was plain that the *Investigator* must be abandoned.

The *Enterprise* meanwhile had followed M'Clure in the spring of 1851, and passed two years in searching every shore and passage she could find, while her men made sledge-journeys far and near, as M'Clure's men were doing, and once came within a few miles of Point Victory, where Franklin's remains would have been found. At last, in the spring of 1854, she succeeded in making her way back along the American coast, and returned to England, completing one of the most remarkable of Arctic voyages.

During their absence the friends of Franklin had not been idle. The apparent sacrifice of this fine character aroused almost or quite as much interest in America as in England, and Yankee shipmasters knew the north as well as did the men of England and Scandinavia. Henry Grinnell, a prominent merchant in New York, furnished the money to fit out two ships, the *Advance* and *Rescue*, commanded by Lieutenants De Haven and Griffiths, of the United States navy. They assisted in the search about Beechey Island, then struck north and discovered Grinnell Land, after which they returned before the winter had closed in. With them was a young physician and traveler, Dr. Elisha Kent Kane, who persuaded Mr. Grinnell to send him again to the north, less to search for Franklin, whom he had despaired of, than to prosecute explorations in higher latitudes. In 1853, in command of the little brig *Advance*, manned principally by whaling men, he left New London, Conn., and made his way straight up to the head of Baffin's Bay, which narrows northward into Smith Sound, where, on the eastern, or Greenland, shore of its expansion, since called Kane Basin, he was stopped by ice and remained a prisoner until rescued in 1855.

Dr. Kane wrote the histories of these expeditions, and especially of the latter one, in books so charmingly expressed, and abounding in such novel information, that they were read like romances in every home in the land, and did more to fire the ardor for Arctic discovery which has ever since glowed in this country, than anything else that had been said or done. The most immediate result was that Dr. I. I. Hayes, who had been with Kane, took a ship to Smith Sound and spent the winter of 1860-61 there, but with little result. More came from the expeditions led by an enthusiastic journalist of Cincinnati, Charles F. Hall, but before speaking of these, let us return to the English search for Franklin.

Undeterred by the failure of Austin and Penny, or the silence of Collinson and M‘Clure, the British government in 1852 despatched again the four vessels used by Austin, and added a fifth, the *Assistance*, and a store-ship, the *North Star*, to form a depôt of supplies at Beechey Island. The old haphazard ways had given place to very systematic methods of advance and rescue; but steam was little employed as yet, because of the trouble and cost of supplying coal, although two small steam vessels, as tenders, accompanied this, the largest and most bountifully equipped expedition that had yet started out. The fleet, under command of Sir Edward Belcher, proceeded through Lancaster Sound, beyond which they scattered somewhat, and spent the first winter in extensive sledge-journeys, during which they discovered (by a message that M‘Clure had left on Melville Island) where the *Investigator* was imprisoned, and rescued all its people in June, 1853.

This great expedition learned nothing of Franklin, although it did learn much of other Arctic matters, and left the map substantially complete south and west of Jones Sound; but its honors rested upon M‘Clure, who, first of all recorded men, had really made the Northwest Passage by sailing and sledging around the northern end of America. The settlement of this long-discussed matter had proved it of no practical value; but the British Parliament kept its word, and gave £10,000 (half of the promised reward) to the officers and crew of the *Investigator*, besides raising M‘Clure to knighthood. An incident of this expedition is the fact that Kellett's abandoned ship *Resolute* survived crushing long enough to drift out through Barrow Strait and Lancaster Sound and down into Davis Strait, where in September, 1855, she was found and towed home by an American whaler. As she was little injured, she was presented to the British government with the compliments of the United States, and a few years later, when she came to be broken up, a fine table was made from her oaken timbers, and returned as a present to Uncle Sam; and it now stands in the private office of the President of the United States in the Executive Mansion at Washington.

Two great facts had now been ascertained. One was that none of Franklin's men or ships survived. The other fact was, that although there was plenty of water north of the American continent, it was so obstructed by permanent ice that probably no vessel could ever make its way through from the Atlantic to the Pacific; none has done so yet, despite the determined effort of the steam yacht *Pandora* in 1875, but ships from the east have reached points also reached by ships from the west. The everlasting ice sheet of the polar ocean, ever crowding down upon this northern coast and into the channels between the islands north of it, forms a barrier that will very rarely, if ever, pause or open long enough to let a vessel through, even south of King William and Victoria lands. The outflowing warm waters of the rivers or other influences may sometimes produce a narrow space comparatively free from ice in summer along the shore of the continent and greater islands; but everywhere off shore, and never at a great distance, begins a thick mass of perpetual ice, which, it is believed, extends across the pole like a cap, and reaches on the other side nearly to Petermannland. To this has been given the name of the Paleocrystic Sea, or sea of ancient ice, and nothing is known of it beyond the blue cliffs of its margin that confronts the explorer as he gazes abroad from the hills of the Parry Islands or Banks Land, or vainly seeks in some lone vessel north of Alaska or Siberia to penetrate its glassy front.

So thoroughly were the islands of this archipelago explored, and so unpromising seems further study, that Arctic voyagers have long ceased to risk their ships there, and the story of Franklin's fate was finally learned by land travelers. As early as 1854 Dr. Rae and a party of Hudson's Bay Company's men had traveled over land and ice to King William's Land, proved it an island, and heard stories of the death by famine and cold of white men who could be no other than the Franklin crew, as was further shown by various relics which Dr. Rae obtained from the Eskimos. Dr. Rae claimed and received £10,000 of the reward offered by the British government. The next year another party, going down the Great Fish River, recovered many other articles from Eskimos at the mouth of the river and on Montreal Island. It was evident even then that every one had perished in an attempt, nearly successful, to reach the mainland at the mouth of this river. Lady Franklin, however, despatched an expedition in the *Fox*, under the command of the experienced M'Clintock, which at last brought back, not her husband, but the satisfaction of knowing fully his fate.

All along the west and south coast remains of articles belonging to the ships were found, and skeletons — two of them in a broken boat; and finally in a stone cairn a written record that briefly told the tale of disaster.

In 1845–46 Franklin quartered at Beechey Island, on the southeast coast of North Devon, after having ascended Wellington Channel to latitude 77°. and returned west of Cornwallis Island, which was an exceedingly successful season's work. In the autumn of 1846 he had turned toward the south, but had been stopped by and frozen into the masses of ice that come ceaselessly down M'Clintock Channel and press upon King William's Land. Had he known King William's Land to be really an island he need not have exposed himself to this. During all the summer of 1847 the ships remained firm in their icy bonds. Sir John Franklin died, and Captain Crozier took command. The spring of 1848 brought no hope, and

in April the ships were abandoned. The crews started southward along
the shore, dragging two boats (one of which was soon abandoned) and
many sledges. The Eskimos said the men dropped down one at a time,
from weakness and hunger; but it is believed that many of them were killed
by the savages for the sake of what few things they had with them —
precious articles to those natives. It appears that one of the vessels must

WORKING THROUGH AN ICE-FLOE, IN TOW OF A BERG.

have been crushed in the ice, and the other stranded on the shore of King
William's Land, where it lay for years, forming a mine of wealth for the
neighboring Eskimos. Some years later Lieutenant Schwatka and W. H.
Gilder, traveling with Eskimo parties in the region near the mouth of the
Great Fish River, found the graves of the last remnant of the party, and
recovered still other relics of this dreadful calamity. Let me copy for you
here the postscript, written by Crozier and Fitzjames, to the short record of
their work. It is startlingly brief and impressive:

April 25, 1848. H. M. ships *Terror* and *Erebus* were deserted on 22nd April, five leagues
N. N. W. of this [Point Victory], having been beset since 12th September, 1846. The officers

and crews, consisting of 105 souls under the command of Captain F. R. M. Crozier, landed here in lat. 69° 37′ 42″ N., long. 98° 41′ W. Sir John Franklin died on the 11th June, 1847; and the total loss by deaths in the expedition has been to this date 9 officers and 15 men. We start on to-morrow, 26th April, 1848, for Back's Fish River.

It would be tedious to attempt to chronicle the almost yearly excursions into the north, but a few ought to be spoken of. One such has been alluded to — that of Charles Hall, a Cincinnati journalist, — who enlisted the aid of the American Geographical Society, and then prepared himself by going upon a whaler and spending the winters of 1860–61 and 1861–1862 among the Eskimos near Cumberland Sound, where he found the remains of a stone house built by Frobisher in 1578. Again, from 1864 to 1869 he was living with the wandering Eskimo north of Hudson's Bay, preparing himself to undertake an expedition which may be said to be the first whose avowed object was to try to reach the North Pole. The United States government furnished him the steamer *Polaris*, and a small but efficient body of scientific assistants, one of whom was Emil Bessels. The *Polaris* passed through Smith Sound, and after completing the exploration of Kennedy Channel, and discovering that beyond its expansion into Hall Sound it continued straight northeastward, forming Robeson Channel, Hall stopped his ship and by sledge-journeys reached Cape Brevoort, above 82° N., whence he could see the open polar sea. This was not only far beyond any previous northing, but his work added immensely to our knowledge of both Grinnell Land and northwestern Greenland, and prepared the way for further successes.

This sledge-journey was, however, too great a strain, for he had hardly returned to his ship when he sickened and died. The next season (1872) Dr. Bessels and Sergeant Mayer reached on foot 82° 09′ N., a few miles farther than Hall. This accomplished, an attempt was made to return, but the steamer was soon inclosed in the pack, and drifted helplessly southward for two months, until off Northumberland Island, when a violent gale loosened the pack and nearly destroyed her.

At length the danger became so great that on October 15th boats and provisions were put on the ice, on which nineteen of the crew had disembarked. Suddenly the ship broke away, and the party on the ice drifted slowly 193 days, and were picked up off the coast of Labrador, in 53° 35′ N., by a whaling steamer 1,300 miles from where they had parted with the *Polaris*. The party in the ship reached Littleton's Island, where they passed the winter, building two boats from the boards of the vessel, in which they set sail southwards in June, 1873. On the 23d of that month they were picked up by a Dundee whaler, and ultimately reached home.

Only three years before that a very similar experience had happened to the smaller ship of a German expedition under Captain Koldewey, of which the larger went up the east coast of Greenland to 75½° N., where a grim headland was named Cape Bismarck. It is just south of the land sighted by Lambert in 1670. The little *Hansa*, however, was crushed in the ice near Scoresby Sound. The crew escaped to the floe, where they built a house of blocks of patent fuel, filled it with provisions, and trusted themselves to the great Arctic current which carried them south, at the rate of about sixty-five miles a day at first, until finally, in June, 1870, it took them to the Moravian missions near Cape Farewell, more than twelve hundred miles from where they were wrecked.

The seas and archipelagoes north of Europe were being questioned, all this time, as well as those north of America. The Norwegian fishermen had been familiar with Spitzbergen waters from long ago, but it was not until 1863 that the group was circumnavigated. The next year Captain Tobieson sailed around Northeast Land, and in 1870 Nova Zembla was circumnavigated, and the mouth of the Obi reached.

The men who did these feats were sealers or shark-fishers in small stanch Norwegian schooners, which flocked in Barentz Sea at this period, and they furnished invaluable material, as did the whalers and sealers of American and Scotch ports, for the ice-pilots and crews of the scientific expeditions which now began to go to the north: moreover many of the commanders were trained by amateur service in such vessels. It was thus Nordenskjöld began his experiences in 1864. Among these earlier expeditions was an Austrian naval lieutenant, Julius von Payer, who became notable, not only because he interested a new nation in Arctic research, but because of his discoveries. His first experience was with the German expedition to Greenland in 1869, and in 1871 he and another Austrian navy officer named Weyprecht spent the summer in examining the edge of the ice between Spitzbergen and Nova Zembla.

Their observations led them to project an expedition to try again at that place to penetrate eastward, and effect the Northeast Passage, which had been regarded as hopeless for the past hundred years. The idea of making an Austro-Hungarian expedition of it aroused great enthusiasm in that empire, and Payer and Weyprecht were furnished with the large steamer *Tegethoff*, equipped as well as possible, with Weyprecht in command, while von Payer was to lead all sledge-parties. She reached the northern end of Nova Zembla in time to get into comfortable winter quarters, but instead of escaping in the spring was kept imprisoned in the ice, drifting steadily northward before the prevailing wind until, in October, land was approached, near which the ship again became a fixture for the winter of 1873-74. In March Payer began to make exploratory journeys,

A SUMMER SCENE OFF NOVA ZEMBLA

and found that they had discovered a group of mountainous islands,
separated by broad and deep channels, which he named Francis Joseph
Land, in honor of the Emperor of Austria-Hungary.

By this time summer was approaching, when it was plain that the
Tegethoff must be abandoned, and an attempt made to get home afoot. On
the 24th of May three boats were placed on sledges, other sledges were
loaded with provisions, and the ship's company started on another one of
those Arctic marches that often end at so sad a goal. Until the 14th of
August they were plodding over the ice before they reached the edge of the

pack and launched their boats, in which they sailed for three weeks before being picked up by a Russian vessel.

This has always been regarded as one of the greatest achievements in polar work of this century, not only because of the heroism and skill shown, and the new lands discovered, but because it promised so much for the future—a promise that has been largely fulfilled.

The next important expedition was another attack upon the Northeast Passage, the hope of which would not "down"; and it was under the leadership of Professor Adolf Erik Nordenskjöld, a Swedish geologist and naturalist of Stockholm, although born in Finland, who had made several previous journeys to Greenland, Spitzbergen, etc., which were fruitful of scientific results. Then he turned his attention to Siberia; and in 1875 and again in 1876 he sailed to the mouth of the Yenisei, as also Captain Wiggins of Sunderland, England, was then doing, in a profitable trade with the Siberians, which has been kept up more or less regularly ever since. These experiences convinced him that it was worth while to try once more to work one's way through the Siberian ocean to Bering Strait.

He obtained and outfitted the steamer *Vega*, and arranged that a smaller supply-steamer, the *Lena*, should accompany him as far as the mouth of the river Lena—a bold proposition in itself, for that was a thousand miles beyond the Yenisei. Nevertheless, this program was carried out; for leaving Gothenberg on July 4, 1878, a month later they were traversing the Kara Sea, and on August 19 passed Cape Chelyuskin, which, up to that time, had defied all attempts and has since closed the gate to all but the daring Nansen. A week later the mouth of the Lena was reached, and the little tender, unloading her coal and other stores into the depleted hold of the *Vega*, turned west, and actually sailed back to civilization uninjured.

The *Vega* then hastened on eastward, and came near getting right through to Bering Strait in that one season; but this was more than the indulgent Arctic gods could grant, and at the end of September the men found themselves frozen into the ice off North Cape (where Cook turned back in 1778), only one hundred and twenty miles from Bering Strait. Here they were near shore, the country was inhabited by Tchuktches—a nomadic people, with herds of reindeer, who take the place in Siberia of the Eskimos of Arctic America; and the time was well spent in gathering a knowledge of these people and their country, and in making very valuable collections in zoölogy and anthropology.

It was not until July 18, 1879, however, that their prison-gates opened, and the *Vega* steamed on. These waters were familiar enough to navigators; and Nordenskjöld proceeded straight east, passed down through

Bering Strait on the next day but one (so near was he), and thus easily accomplished that which had baffled men since first it had been tried by the unfortunate Willoughby three hundred and twenty-six years before.

But though the Northeast Passage had thus been found, it was of no more practical value to commerce than the solving of the Northwest Passage had been, and the value received from the cruise was in the scientific information gained, the more accurate delineation of the coast, and the increased knowledge of winds, currents, magnetic phenomena, and the behavior of the floating ice-fields on that side of the polar area. When at last, however, the *Vega* had circumnavigated the globe by this extraordinary course, returning home through the Suez Canal, as no Arctic expedition had ever been expected to do, its commander was made a baron, and all his men were loaded with praises and honors, while his book, "The Voyage of the *Vega*," printed in four or five languages, spread their fame throughout the world.

Now while the *Vega* was drifting slowly about northeast of Siberia during that early summer of 1879, not only were Schwatka hunting for Franklin relics with the Eskimos of King William's Land, the Danish Captain Jansen tracing the northeast coast of Greenland, and Dutch and English explorers investigating the neighborhood of Francis Joseph Land, but within a few leagues of Nordenskjöld and his men there was beginning one of the most dreadful of those tragedies that have seared with suffering the track of Arctic exploration since men began to pry into the secrets of the frozen North: I mean the story of the *Jeannette*.

Many readers of this book will easily remember the intense interest which the starting of this expedition created in the United States, for it was organized at the suggestion and expense of James Gordon Bennett, the proprietor of the New York *Herald*. The government coöperated, however, lending from its navy the officers and men needful, and otherwise aiding the project. The vessel itself was the steam yacht *Pandora*, which had been proved a worthy craft by Sir Allen Young in his search for the magnetic pole in 1875, and which Mr. Bennett had bought and rechristened.

Supplied with everything science and experience could suggest, the *Jeannette* sailed from San Francisco on July 8, 1879, and missing the incoming *Vega* among the fogs of Bering Sea, passed through into the Siberian ocean, bound poleward. The last report of her was that she had been seen September 3d steaming toward Wrangell Land, which had been sighted by American whalers in 1867, and was generally regarded as of continental extent northward. It is now known that De Long intended to reach it and winter there; but to his dismay he could not escape from the ice-pack, and to his astonishment found himself drifting past the northern margin of Wrangell Land, thus proving it an island about seventy miles long.

When two years had passed and no tidings had been received, the United States government equipped a search expedition in the steamer *Rodgers*, commanded by Lieutenant Berry, which in 1881 reached and examined Wrangell Land, and then went north farther even than Collinson, reaching 73° 44', the highest point yet attained immediately north of Bering Strait, where the palæocrystic ice spreads much farther from the pole than on the American side. But he found no trace of the *Jeannette*, and himself had a hard time getting home, for the *Rodgers* was burned in her winter quarters.

What then had befallen the lost vessel? She had become beset in the ice and drifted with the pack around the north end of Wrangell Island, and then west, until at the end of twenty-two months she had been crushed, and sunk on June 12, 1881, in latitude 77° 15' N., and longitude 155 E. Two small islands, named Jeannette and Henrietta, had been visited some distance east of the scene of the catastrophe; but when the crews, saving themselves and what little they could on the ice, started to drag their boats and sledges homeward, they headed directly south, and soon found a new island, named Bennett, which is the northernmost of the New Siberia group.

It would be a sad task, were it possible, to relate here the frightful hardships of that journey through the fast-gathering Arctic night toward the bleak coast of Siberia. Having passed the islands, open water was found, and the starving men embarked in their three boats for the mouth of the Lena; but soon they were separated in a storm, and each one proceeded as best he could. One boat foundered in the first gale. Another, in charge of Melville (now engineer-in-chief, U. S. N.), reached an eastern mouth of the river and ascended it to a Russian village. A third boat, with De Long and others, also reached the Lena delta, but only two seamen were able to proceed afoot to Bulun, a far-away Russian settlement. Melville heard of this, and made haste to start out searching parties, but they were too late. De Long and his crew had died of exhaustion, and it was not until the next season that their bodies and records were fully recovered.

Nevertheless, as we are assured by experts, the results of this unfortunate expedition were important, physically and geographically. "They covered some 50,000 square miles of polar ocean, and clearly indicate the conditions of an equal area between their line of drift and the Asiatic coast." De Long believed the Siberian ocean to be a shallow sea, dotted with islands; and his conclusions have been confirmed by the admirable scientific work since of Toll, Bunge, and other Europeans who have explored the Liachoff Islands and other places in that part of the Arctic realm.

The desire for scientific study of the polar world had now become the motive for northern research, though men were still ambitious to reach the pole; and when Sir George Nares returned from the great British expedition of 1875, to tell how the men of the *Alert* had reached a wintering-point beyond Robeson Channel, on the west coast of Greenland, in latitude 82° 27′ N., and that Markham and a sledge-party had gone about one degree farther (to 83° 20′ 26″ N.), greater pride was felt in this fact, perhaps, than in the careful observations and collections that the ships had made. This remained the advance record until the memorable feat of Lieutenant Lockwood of the American Greely expedition eight years later.

This expedition was one of several acting in concert, according to a scheme suggested by Weyprecht, and perfected at international congresses of interested men meeting at Hamburg in 1879 and at St. Petersburg in 1882. This plan was for the establishment by various governments of a ring of stations as far within the Arctic circle as practicable, where simultaneous daily observations of the weather, magnetic conditions, tides, currents, etc., might be made. The arrangement was begun in the summer of 1883, and observing stations were established by Austria on Jan Mayen Island; by Denmark at Godthaab, Greenland; by Germany on Cumberland Bay, west of Davis Straits; by Great Britain at Great Slave Lake, Canada; by Holland at the mouth of the Yenisei; by Norway at Alten Fjord, Norway; by Russia at the mouth of the Lena, and on Nova Zembla; by Sweden on Spitzbergen; and by the United States at Point Barrow, Alaska, and, farthest north of all, Lady Franklin Bay, Greenland. Nothing need be said about most of these stations—all were successful except the Dutch; but to the last-named belongs a story that Americans will not forget.

The command of the Lady Franklin Bay Station was assigned to Lieutenant A. W. Greely—not a naval lieutenant, but, like Schwatka, a cavalry officer, then assigned to duty in the Signal Service, to which (because it then supervised the Weather Bureau) the government had intrusted this matter. A steamer easily conveyed Greely and his party to Lady Franklin Bay, and left them there with a good house ready to be set up, and supplies of all sorts for two years. The prescribed series of observations with barometers and thermometers, wind-gages, tide-gages, magnetic instruments and all the rest, were at once begun, and two winters passed comfortably enough. Dogs and Eskimo drivers had been obtained, and several journeys were made, of which the most important was Lockwood's advance toward the pole, of which an account has been succinctly supplied by General Greely himself in his admirable "Handbook of Arctic Discoveries."

Lieutenant J. B. Lockwood, one of the principal assistants, who had already displayed great skill and energy in sledging, even in prolonged temperature as low as 81° F. below freezing, undertook a long exploring trip up the Greenland coast, to or beyond Cape Britannia. A large party went with him at first, but gradually men were sent back, after establishing supply-depots. "The journey onward was marked by severe storms, rough ice, broken sledges, snow-blindness, minor injuries, and—worst of all for loaded sledges—soft, deep snow." At last, some distance north of Cape

SCENERY OF GRINNELL LAND AND THE ARCTIC SEA.

Bryant, all turned back except Lockwood, Sergeant Brainard, and an Eskimo, Christiansen, who, with twenty-five days' rations, pushed on. In five and one half days they had reached Cape Britannia—the farthest north of the Nares expedition—82° 20′ N. Halting here only long enough to study the landscape from its summit, and make sure of the remarkable fact that this northern end of Greenland is free from the ice-cap, whose northern limit is about lat. 82° N., they rounded a cape, and crossing channel after channel filled with ice, which showed that all this district is an archipelago, reached on May 10th Mary Murray Island, 83° 19′ N. "A violent gale delayed them sixty-three hours, the cold exhausting them physically and the delay mentally. If weather forbade travel, life must be sustained; but they tasted insufficient food only at intervals of fifteen, twenty-four, and nineteen hours — the last as clearing weather made progress possible. Floes so high that the sledge was lowered by dog-traces, ice so broken that the ax cleared the way, and widening water-cracks in increasing numbers impeded progress. But, despite all obstacles, they reached, May 13, 1882, Lockwood Island, $83° 24'$ N., $42°, 45'$ W., the farthest of their journey, and the highest north [by land], then or now."

They could see land several miles northeast, which they named Cape Washington, the highest known land, and toward the north could overlook a polar sea to within three hundred and fifty miles of the pole. Even here plants were numerous, and foxes, hares, lemmings, and ptarmigans existed. The three heroic travelers returned safely, reaching headquarters on June 3d. Another expedition by Lockwood and his two companions explored and located the west coast of mountainous and glacier-girt Grinnell Land, where the musk-ox and Eskimo hunters range to the northern border.

The summer of 1883 brought no relief-ship, and the plan of escape must be put into execution at once. A ship had, in fact, tried to reach Greely in 1882, but, failing, had left supplies of provisions at Cape Sabine and elsewhere. In 1883 another relief expedition sent north was dreadfully mismanaged, and finally the ship itself was lost, and, instead of leaving supplies, took away all that had been stored at Cape Sabine — the precise point where they were to be needed.

Leaving Lady Franklin Bay in August in open boats, the party managed, after desperate exertions, to get near Cape Sabine, and safely landed on Bedford Pim Island, on the northwestern shore of Smith Sound, October 15, 1883. Of the misery that followed, let Greely himself tell us:

Winter had begun, the polar night was imminent, clothing in rags, fuel wanting, and forty days' rations must tide over 250 days, till help could come. The main party put up a hut of rocks, canvas, boat- and snow-slabs, while selected men scoured the coasts for caches, sought land-game, and watched seal-holes, until utter darkness drove all to the hut. Scientific observations were unremittingly made, amusements devised, a spring campaign planned, and the returning sun found only one dead. Efforts to cross Smith Sound failed, and a hunting trip to the west found a new (Schley) land, but no game. Finally game came so inadequately that food failed, and one by one men died — Jens seal-hunting, and Rice striving to bring in a cache. Courage and solidarity continued; and if Greely gave to the maimed Ellison double food while it lasted, he did not hesitate to order in writing the execution of a man serving under an assumed name of Henry, who repeatedly stole sealskin thongs, the only remaining food. Flowers, plants, seaweed, and lichens eked out life for the six, till June 22, 1884, when the relief-ships, *Thetis* and *Bear*, under Captain W. S. Schley and Commander W. H. Emory, rescued them. Records, instruments, and collections were saved to tell the story of an expedition that failed not in aught intrusted to it, and whose members perished through others.

To another piece of brilliant work, that of Lieutenant R. E. Peary, U. S. N., I can give only a few words, because, like so much else that might be said of Arctic researches, it was by land rather than by sea. By extraordinary courage, skill, and endurance, he twice crossed northern Greenland, showed that it is an island having a northern shore free from inland ice in about 82° north latitude, and made stronger Greely's conclusion that the lands visited and seen by Lockwood, north of Cape Britannia, are detached islands. Peary's work may be said to have completed the map of the continental boundary of the Arctic Ocean, but he is still busy there.

Of Nansen, on the contrary, I ought to say as much as I can, because his extraordinary voyage in the *Fram* was perhaps more purely an examination of the Arctic *Sea* than any other ever made. Dr. Fridtjof Nansen was a young Norwegian who had already made his mark in Greenland, where, soon after 1880, articles began to be found that had belonged to the *Jeannette*, and apparently must have drifted thence from where she was lost off Siberia. This was only a part of the indications that convinced Dr. Nansen that a current flowed across the unknown polar space from the neighborhood of Alaska to the northeast coast of Greenland, and thence became the great Arctic current that we recognize south of Iceland. He argued that if a vessel could find this current north of eastern Siberia, she would be moved with it until she emerged into the Atlantic. Incidentally she might drift directly over the pole.

With this in view, he raised funds to build and equip a small wooden vessel, furnished with both steam and sails, which was so shaped by the roundness of her bottom, and so amazingly braced and strengthened within, that before any "nips" of the ice would crush her, the pressure would lift her out of water — as, in fact, happened many times in the course of her wonderful excursion. Nansen chose twelve companions,[1] and though some of them were educated men of science, others skilful sea-captains, and others common sailors, all lived and worked together in one cabin as brothers — the happiest and healthiest lot of men that ever ventured into the hyperborean kingdom of desolation.

Leaving Norway in July, 1893, he struggled through the Kara Sea, and it was not until late in September, 1894, that he found himself permanently frozen into the great polar pack, north of the New Siberian Islands; but even then he was neither so far north nor so far west as he hoped to get, and feared that he was south of his supposed current. For the story of the strange life led by those thirteen men on that drifting ship, safe, abundantly

[1] The success of this most hazardous venture, although its crew numbered *thirteen*, is equal to the success of Columbus's first voyage, although it began on *Friday!* "Luck" has no show when it is pitted against pluck.

provisioned, dry, warm, lighted by electricity (power for the dynamos being gained by a windmill), I can only refer you to Dr. Nansen's book, "Farthest North," one of the most interesting Arctic volumes ever penned. Turning, zigzagging, now advancing and again retreating as the constantly moving ice swayed here and there under the pressure of wind or the dragging of currents, they nevertheless made a gradual progress westward.

By March they had reached a point near the crossing of the 70th meridian and 85th parallel, and were still fixed in the ice. Then Nansen, taking with him Lieutenant Johansen, started north by dog-sledges, in an attempt to reach the pole. They could take very few supplies of any sort, and how far north they would be able to travel must depend upon their ability to return, not to the *Fram*, which would drift on, but to the islands of Francis Joseph Land, far away south. The ice, bad at first, grew worse as they proceeded, being one long stretch of hummocks and jagged ditches, with now and then a lane of open water around which they would toil in misery only to find a worse one ahead. On April 7th it became certain that they must turn back. This was "farthest north," indeed — just above the 86th degree, hardly 275 miles from the North Pole. Then it was a race against death by cold, or drowning, or starvation. One by one the dogs were killed to furnish food for the remainder. At last, after almost superhuman labors and thrilling escapes from freezing and drowning and the attacks of famished bears, they reached Francis Joseph Land, and spent a winter in a hut made out of stones, earth, and raw walrus hides. The next spring they plodded on, and by good chance found the camp of the Jackson-Harmsworth surveying party (which a few days later would have gone away in its steamer), by whom Nansen and Johansen were carried to Norway in August, 1896.

A week later the *Fram* came in, with every one well and hearty, having emerged from the ice just northwest of Spitzbergen.

Since Nansen's return another Scandinavian, S. A. Andrée, with two companions, has disappeared into this same desert of ice and silence, in a balloon carrying a boat, sledge, tent, and various supplies. It was his intention to reach the pole if possible, and to do whatever else circumstances permitted. Since his departure, on July 10, 1897, from Spitzbergen, he has not been heard from, except by a pigeon-message two days later.

THE SOUTH POLE

We have followed up to date the history of adventurous and scientific exploration of the hardly yielding, yet steadily narrowed, circles of unknown

A PENGUIN-ROOST ON THE BEACHES OF VICTORIA LAND.
Drawn by the Antarctic explorer Borchgrevink.

coasts and waters about the North Pole. Let us now see what, thus far,
has been done to wrest from the ocean and ice of its Antarctic antipodes the
secrets of the South Pole.

Almost three hundred years ago the existence of islands far to the southward of any continents became known to navigators, who were driven
thither by bad weather, and little by little was added to the map of this
desolate region ; but it was not until 1772 that any one went into that
terrible Antarctic sea for the express purpose of a survey. This man was
the intrepid Captain Cook, and though he sailed a third of the way around
the globe in his efforts to find an entrance through the icy barrier, he could
never penetrate beyond 71° south latitude, which is equal to North Cape,
or the town of Upernavik, in the Arctic region. Later captains did little
better, until 1841, when Sir James Ross, in his ships *Erebus* and *Terror*,—
the same vessels which afterward met their destruction with the ill-fated
Franklin expedition,— skirted the edge of the thick ice that everywhere
clothed the land, though it was midsummer, and finally reached the base of
the southernmost land yet known on the globe — a magnificent mountain-
chain stretching away to the south from latitude 78° 10'.

The most conspicuous point of all this range of polar mountains, which

rises from an unexplored continent or great island called Victoria Land,
is the volcano Mt. Erebus. It was in eruption at the time of Ross's visit,
and the explorer tries to tell us of the splendor of its display when the wide
glistening waste of snow and the deep blue of the ocean and the starry sky
are lit up by the column of fire hurled thousands of feet heavenward from
its crater: but who can picture the grandeur of such a scene! This volcano
is about 12,400 feet high, and an extinct neighbor, Mt. Terror, is still higher;
while a third peak, Mt. Melbourne, exceeds 15,000 feet in altitude, and
like all the rest is covered with everlasting snow and glaciers from the
tempestuous water's edge to its lonely crest.

Meager as this information is, it is about all we know of the surface of
the globe within the Antarctic circle; and it will be extremely difficult to
learn much more. In a latitude much farther from the pole than that where
in the north vegetation is abundant, and men and animals live all the year
round, the severity of the Antarctic climate cuts off all life, and constantly
seals the water under a cap of ice. The coasts and outlying islands thus
far examined appear to be wholly volcanic, often composed of nothing but
alternate layers of ashes and ice; but the *Challenger* staff dredged up from
the edge of the ice south of the middle of the Indian Ocean pieces of gran-
ite-like and other rocks, such as belong to land regularly formed; so that
probably the whole uplift does not consist of volcanic materials; and, furthermore, rocks containing fossil plants have been found on some of the southernmost islands which show that in past ages — the period of the coal
deposits — the climate of that end of the world was mild enough to support
forests of trees and, doubtless, a large variety of herbage and animals.
Now most of the coast is unapproachable on account of a border of sea-ice,
or else cliffs of moving land-ice (glaciers) that give off the flat, table-topped
icebergs characteristic of the south polar waters. No trace of any land
animal — except visiting sea-fowl — has been found, and only a little of the
simplest plants (lichens); nor is this surprising when we learn that the highest noonday heat of summer is only a little above the freezing-point.

Why this intense cold and dreadful desolation exists so much farther
from the pole in the southern than in the northern hemisphere, I need
hardly explain to you; for you will recall that in the north the continents
are so broad as to form almost an unbroken wall about the narrow polar
sea, confining its cold waters, warming the air by wide radiation, and guid-
ing the heated flood of the Gulf Stream straight into the northern sea. In
the southern hemisphere, on the other hand, an immense breadth of ocean
south of latitude 40° is broken by no land of any account, and the southward flowing warm water from the equator becomes spread out so thin

upon the vast surface that it is rapidly chilled. It is now generally believed, as has been hinted, that the south polar region is a continental mass, deeply buried in an ice-sheet that is ever fed in the center as fast as it wastes away at the circumference; for the prevailing winds there tend toward the pole from all sides, and carry loads of moisture to be condensed and fall in ceaseless snows.

The Antarctic seas, however, are by no means lifeless, but abound not only in fishes,—cod are said to throng in these waters in prodigious num-

ICE-CLIFFS AND TABLE-TOPPED BERGS, CHARACTERISTIC OF
THE ANTARCTIC REGION.

bers,— but several varieties of whales, dolphins, and their kin (which will be described in one of the later chapters), and many kinds of seals, notably the huge sea-elephant, now becoming rare elsewhere. Then, too, the Antarctic islands and headlands are the resort of enormous flocks of certain sea-birds, all different from the Arctic species of their families, which subsist upon the fishes and less creatures in the water, and go to the lonely shores outside the ice-cap only for rest and to make their nests. Of all these the penguins are most numerous and most hardy, and a whole chapter might easily be given to their quaint appearance and quainter ways. It also appears probable that certain migratory birds — especially beach-feeding kinds — regularly visit the Antarctic continent in summer from Patagonia, and breed there.

Now what has been gained by all the expense, exertion, and hardship of polar exploration? What has been the charm that has led wise and brave men to overcome terrific obstacles, and turn again with deeper and deeper longings toward the mystic icy regions? Lieutenant Maury has given one answer: "There icebergs are framed and glaciers launched. There the tides have their cradle; the whales their nursery. There the winds complete their circuits, and the currents of the sea their round in the wonderful system of interoceanic circulation. There the Aurora Borealis is lighted up, and the trembling needle brought to rest; and there, too, in the mazes of that mystic circle, terrestrial forces of occult power and vast influence upon the well-being of man are continually at play. . . . Noble daring has made Arctic ice and waters classic ground. It is no feverish excitement nor vain ambition that leads man there. It is a higher feeling, a holier motive, a desire to look into the works of creation, to comprehend the economy of our planet, and to grow wiser and better by the knowledge."

To polar explorers we owe not only the discovery of the waters, coasts, and archipelagoes that now are accurately outlined upon our maps within the Arctic and Antarctic circles, but vast and valuable products — whale-fisheries, seal-fisheries, cod-fisheries, and many other additions to the wealth of the world from the sea, while the Arctic lands have yielded furs and other valuable things in great quantity. The study of the people living under those adverse northern conditions has been highly instructive, assisting us to reconstruct the life in the primitive world; and what we have learned from the records of the Arctic rocks has thrown a bright and unexpected light upon the antiquity of the globe.

To studies of the ocean and atmosphere in very high latitudes science is largely indebted for new facts in magnetism, in the movements of the air and causes of climate, in the formation and behavior of ice and icebergs, in the action of tides and ocean-currents, and in many other departments of knowledge, all of which have been made of use especially to the navigator. Nor has this cost over much. Attention has been called to every casualty, and the romantic light of adventure has brought into high relief all the hardships and sometimes horrors of Arctic experience; but the records show that the average of loss and suffering in Arctic work is not greater than that of ordinary seafaring and naval careers. Sir Leopold M'Clintock has stated publicly that during the thirty-six years when Great Britain was most active in polar research, she lost only one expedition and 128 persons out of forty-two successive expeditions sent out, and never lost a sledge-party out of a hundred that made overland journeys.

After all, no doubt, the best result has been the human heroism displayed, and the human sympathy developed. "There are," exclaims Professor Nourse, "and ever will be, fair fruits born out of such acts of high aspiration, energy, and fortitude, in those who have gone out, and in their liberal supporters; exemplars for the lifting up of the discouraged, the education of the young. Certainly volunteers for the paths of discovery will offer themselves until the fullest additions to the domain of science have had their ingathering."

EAGER TO BE FIRST ASHORE IN A NEW LAND.

DRAWN BY HOWARD PYLE   ENGRAVED BY M. HAIDER.

THE "CONSTITUTION'S" LAST FIGHT.

CHAPTER VI

WAR-SHIPS AND NAVAL BATTLES

AVAL warfare, properly speaking, begins with the battle of Salamis, 480 B. C., when the Greek fleet, under the guidance of Themistocles, destroyed or put to flight a horde of twelve hundred Persian vessels, and saved Athens, to become the foundation of a strong nation.

Of these ships at Salamis we know very little, except that they were large, open, or partly open, rowboats, having platforms at the stern and prow, and perhaps amidships in some cases, where soldiers might stand and discharge their arrows out of the way of the rowers beneath them, or leap aboard the enemy's boats whenever they could be reached. They were, in short, early types of the galleys which subsequently became vessels of war as powerful and serviceable, under the conditions they were intended to meet, as are our battle-ships to-day, and probably safer as a fighting-place for their crews.

That from rowboats rather than from sail-boats should have been developed the highest type of Mediterranean war-vessel of ancient times is not surprising when one remembers the light and variable winds of that region, the usually smooth seas, the abundance of harbors, and, above all, the need of having the vessels under complete control when all fighting had to be done at short range — chiefly by ramming and boarding, in fact. It must be remembered, too, that labor was cheap; and it was considered that the most proper and economical — not to say humane — use to which prisoners of war could be put was to make them rowers in public ships, while enough remained to be sold as slaves to the owners of private yachts and privateering galleys. One may imagine a worse fate than this.

The earliest war-vessels of the eastern Mediterranean — those of Homer's time, for instance — seem to have been long and rather narrow rowboats, the best of which had two tiers of oars, one above the other, the lower, shorter tier working through oval holes in the side, and the upper in notches or thole-pins on the gunwale. This left the upper rowers exposed, and hence such vessels were called *aphract*, or "unfenced"; and it was not until the Greeks began to become prominent that the bulwarks were raised high enough to protect all the rowers, and war-vessels generally became *cataphract*, or "fenced."

It appears that in very early times war-ships (*biremes*) with not only two tiers or banks of oars, but even those (*triremes*) with three banks, were used; and the trireme became the type of the most numerous and effective vessels of the Greek and Roman navies in their prime. And as weight and power gradually increased, the crushing power of collision began to be utilized, and ramming came in as a more and more important feature in naval tactics. As the Greeks seem to have first applied these new ideas, it is quite likely that their success at Salamis was due to these improvements. The arrangement was this:

From the side of the vessel (inside) projected three rows of benches, a yard apart, horizontally supported at their inner ends by timbers that slanted toward the stern at such an angle that the top seat of each row was exactly above the bottom seat of the row behind it. The oars of the top tier (*thranite*) were about fourteen feet long, those of the middle tier (*zygite*) about ten and one half feet, and the lowermost one (*thalamite*) seven and one half feet. Each oar was so nearly balanced in its oar-port as to work in the easiest manner, tied there by a thong and surrounded by a loose sleeve of leather which kept out the water. Each one of the lowermost oars was worked by a single man, the middle ones by two, and those of the third tier by three or four, as they were of great length.

In later times larger vessels were invented for special purposes — four-banked (*quadriremes*), five-banked (*quinquiremes*), and so on, even up to one of forty banks; but as we are unable to understand how it was possible for more than five or six tiers of oars to be operated, we may leave these extraordinary galleys to special students.[1]

The structure of these vessels gave them the greatest strength combined with lightness. They had very strong keels and stems, the latter peculiarly braced; and along their sides ran waling-pieces, or fore-and-aft bracing timbers, the lowermost curving inward forward, until they met in front of the stem at the water-line, where they were braced by massive tim-

---

[1] An example of the so-called forty-bank galley is illustrated, so far as its forward end will show it, in the picture of the ship of Ptolemy Philopator, on page 43. The forty "banks" appear to be groups of oars in a few tiers.

bers, and prolonged into a sharp three-toothed spur, of which the middle tooth was the longest, reaching out perhaps ten feet. This was covered with metal, usually bronze, and formed the *beak*.

"Above it, but projecting less beyond the stem-post, was the *proembolion*, or second beak, in which the prolongation of the upper set of waling-pieces met. This was generally fashioned into the figure of a ram's head, also covered with metal. . . . These bosses, when a vessel was rammed, completed the work of destruction begun by the sharp beak at the water-level, giving a racking blow which caused it to heel over and so eased it off the beak, releasing the latter before the weight of the sinking vessel could come upon it."

HAMILCAR'S "STAIRWAY OF THE GALLEYS" AT CARTHAGE.

The stem was often carried up into a curving ornament called the *acrostolion*, beneath which was a stout-walled deck-space for sailors or the fighting-men to do their work; and the stern-post similarly supported a lofty, richly ornamented structure (*aplustron*), arching over the officers' quarters.

Platforms extended up and down the center of the ship between the rowers; and over their heads was a deck having walls or bulwarks where the fighting-men and their various "engines" stood. In addition to this an external defended gallery for soldiers and boarders usually ran along the outside of the bulwarks above the oars; and awnings of rawhide were stretched over all to ward off grappling-irons.

It must not be forgotten, however, that these galleys also had three pole-masts, and certain sails — probably a huge split lug, with possibly a square topsail on the mainmast, while the fore- and mizzenmasts carried lateens. At the top of each stick was a round, protected cage filled with archers and slingers — the prototype of our "military mast."

Nor are the size and force of these Greek and Roman men-of-war to be despised. The ordinary trireme had a crew of 200 to 225 men in all, 174 of whom were rowers. The space for cabins and stowage must have been little, but this was of small account, since the war-galleys rarely undertook long cruises, their tactics being a rush and a sharp fight, and then a quick return to harbor, where it was the practice to draw the lighter galleys up on shore each night. The transportation of the ships across the isthmus of Corinth was not, then, so astonishing a feat as it is sometimes called.

Rome's experience, however, gained in war and in suppressing the Levantine pirates, taught her to abandon the heavy, many-banked, unwieldy vessels she had at first developed from Greek and Carthaginian models, and to trust to a much lighter, swifter, and more manageable style, with far less upper structure and rigging, and having only two banks of oars. These were called Liburnian galleys. With this change came naturally one of tactics, capture by chase and boarding taking the place of the earlier attempt to crush by ramming and overriding the antagonist.

The armament comprised not only as many soldiers with bows and javelins as could find room in action, but various machines of offense and defense, such as catapults hurling huge stones or marble grape-shot, spear-headed rams or huge knives that could be run out against an enemy's hull or rigging, arrangements for smashing the enemy's decks, caldrons swung at yard-arms, holding burning pitch or oil to be poured upon the foe, and often cranes (*corvi*), provided with grapples that, if one could be made fast, would lift an adversary out of water, and turn him upside down. No more vivid picture of the life in cruise and battle of a Roman man-of-war's man is known to me than that penned by General Lew Wallace in "Ben Hur," but I cannot, of course, transfer all of it to my pages, as I should like to do, and an extract here and there would only spoil the pleasure in store for you in re-reading it all.

Of medieval naval warfare in the Mediterranean, the struggles between the weak "principalities and powers" that followed the decay of Rome and lasted for a dozen centuries, we know very little. There is more obscurity here than even elsewhere in the dim history of the dark ages. It is evident, however, that not much change took place in naval architecture. The Byzantine empire succeeded to Rome as mistress of the seas, and we know that in the ninth century the Byzantine emperors were still building biremes (then called *dromones*) armed with tubes for spouting Greek fire. It should be noted that boats having only a single bank of oars came now to be called galleys; and this is the first and proper use of the word, though popularly it is now (or until recently was) applied to any large many-oared boat.

With the introduction of gunpowder and cannon into naval vessels, the ornamental top-works — a picturesque relic of which remains in the Venetian gondola of to-day — disappeared, as we see when the clear light of history begins to shine on the fleets of Venice and Genoa, when these cities were leaders of the world in navigation. Turkey — the successor of the old Byzantine empire and of the Greek power — was then, as now, the great enemy of the west, but in those days it was aggressive. Its fleets were strong and well manned, and they threatened to cross the Adriatic and fasten the baneful

A COMBAT OF ROMAN GALLEYS (BIREMES).

grasp of the Moslem upon Italy in revenge for the persecution of the Moors in Spain. Perhaps they would have done so had not John of Austria, admiral of the allied navies of Spain, Venice, and Rome, won that great victory in the harbor of Lepanto, near the isthmus of Corinth, which destroyed nearly the whole Turkish fleet, and released fifteen thousand Christian galley slaves. This was in October, 1571, and it saved the West from being overrun by the barbarous East, as exactly fifteen and a half centuries before it had been saved near Actium, a famous promontory on the northwestern coast of Greece, where Octavius defeated the forces of Antony and Cleopatra.

It is doubtful whether the ships that fought in the later battle were much different in either build or rig from those of the earlier conflict, but their decks no more gleamed with men in armor, and in place of catapult, crane, and caldron were cannonades and falconets, arquebuses and hand-grenades. Perhaps, however, they had already taken on more of that long, low shape characterizing later the French and Italian galleys, common enough in Mediterranean ports up to about one hundred years ago, which differed mainly from the ancient ones in their use of much longer oars or sweeps, balanced upon a sort of extended outrigger or shelf projecting from the vessel's side. The galleass of which we hear in the thirteenth and fourteenth centuries was a large war-ship of this style, which foreshadowed the Atlantic ships, to be spoken of presently, in having castellated structures fore and aft, in which were mounted sometimes twenty guns; besides its two or three lateen-rigged masts, it often had thirty-two sweeps on each side, each about forty-five feet long, and handled with a long, slow stroke by five or six men — in France mainly convicts "condemned to the galleys." [1]

Such vessels continued to be used by the Spaniards, Maltese, Italians, and Turks long after they had been abandoned by the French navy, but latterly, after the suppression of piracy, in which they were of especial service, for the conveyance of important personages and occasions of ceremony rather than for practical service; and in the state barge of the Doge of Venice, brought out annually to this day at the ceremony of re-wedding Venice to the Adriatic, we have a magnificent relic of these stately craft.

TYPE OF VENETIAN GALLEY.

But such boats were adapted only to the comparatively calm and simple navigation of the Mediterranean; and although imitated in the similar waters of the eastern Baltic, they never flourished north of Spain. When they gradually disappeared, their successor inside the gates of Gibraltar was the xebec, which began to appear under Arab or Spanish control in the seventeenth century; this was supposed to be able to withstand any weather, and carried from fourteen to twenty-two guns on deck, with small ports for oars between the guns. A picturesque relative was the Portuguese muleta.

The English liked this kind of vessel on account of its strong sailing

---

[1] Three other terms of similar sound need explanation. The *galiot* was a small, fast galley of the Levant. The *gallivat* was a large, swift, two-masted, armed sail-boat used by Malay pirates. The *galleon* was any Spanish ship sailing to and from the Spanish main; hence, especially a treasure-ship.

qualities, but when they took it into their own stormy waters they found
it necessary to raise its sides to fit them for breasting the high seas that roll
in the open Atlantic or are tossed by the contending tides of the English
Channel, and developed out of it a style of
swift and handy vessel called a frigate.

During all these "middle" ages the northern nations had been sailing and fighting on
the sea as well as the southerners. Stories
of sturdy battles have come down in tradition
and in such chronicles as those of Froissart;
but those old conflicts seem to have produced little change in ship-building or armament until the experience and wisdom brought
back by the Crusaders began to spread abroad
even in the half-savage North, and to produce
that revival of learning which by and by was
to make such striking changes in western
Europe; and here the leaders are Englishmen.

FORECASTLE OF THE "GREAT
HARRY" ("GRÂCE DE DIEU").

In those days no national navies, properly speaking, existed in England, France, or northward. When a monarch wished to transport troops
by water to some other land, or make a naval expedition or campaign, he
fitted out the ships that belonged to the crown as the king's personal property, and compelled his subjects to furnish the rest, just as his feudal provinces and cities and lords were expected to equip and bring to his standard
any land forces required. It was to systematize this method somewhat in
England that William the Conqueror "established the Cinque Ports, and
gave them certain privileges on condition of their furnishing 52 ships,
with 24 men in each, for 15 days, in cases of emergency." Now and
then, at first, Englishmen were disposed to resist the "arrest" of ships,
which might easily mean the ruin of their business; and special laws had to
be made to quell this reluctance. Another quaint and significant feature of
that practice was this: In every fleet one or more ships were set apart as
"royal," and either the king or his representatives occupied them with court
ceremony to carry out the fiction of royal dominion over the sea as well as
upon the land. It naturally followed in England that after her navy had
shown its power, and signalized it especially by a brilliant victory over
Spain in 1380, Edward III should have assumed as an additional title
"King of the Seas"—an act which had far-reaching consequences.

During the fifteenth century something like an established navy was
foreshadowed; but it was not until the reign of Henry VII, when, at the

end of the fifteenth century, the whole world was exploring the oceans and awakening to the importance of sea power, that the first vessel, properly called a national war-ship, was built, equipped, manned, and sustained at government expense by England. This was the *Great Harry* — a floating fortress rather than a ship: for, with her towering, overweighted "castles" fore and aft, she was unseaworthy, and came near being sunk by a slight rolling which poured the water into her lower ports.

But a better known "*Great Harry*" was the *Henri Grâce de Dieu*, built by Henry VIII. This king was the real founder of the British navy, providing for it many good ships, dock-yards, trained officers, and regularly enlisted crews. The advantage of this organization and the superiority of English seamanship were demonstrated in the next reign by the defeat of the Spanish Armada.

England was then at war with Spain, and Philip II thought to end the matter by means of the greatest expedition ever heard of. It began to be prepared in 1587 under the title of the Most Fortunate Armada,[1] but an English squadron under Drake attacked the rendezvous at Cadiz, destroyed over one hundred vessels and huge quantities of stores, and then so ravaged the neighboring coasts as to delay Spain's project for a whole season.

In midsummer of 1588, however, after an unlucky start, in which it was driven back by storms, the dreaded Armada appeared in the English Channel, like a close flock of huge birds drifting along the British coast. It consisted of about 130 ships, seven of which exceeded 1000 tons burden, and numerous small craft, and was armed with nearly 3000 cannon. Its commander was the Duke of Medina Sidonia, who was a most incompetent man for the post, and it bore, besides nearly 10,000 sailors and galley-slaves, over 10,000 soldiers; but this naval force was not intended to attack England until after it had ferried over from Belgium the Spanish army of the Duke of Parma.

To such a force as this England opposed a miserably small fleet — only 34 vessels that could be called ships; but she hastily armed as many more smaller ones as she could, amid great fright and excitement, until finally Admiral Howard commanded 80 or 90 ships and boats. There was no deficiency in his men, however, — the pick of English "sea-dogs" was at his call; and among the leaders of the pack were men we have already met elsewhere — Francis Drake, John Hawkins, Martin Frobisher, and others.

What a sight it must have been on that August day as these ships, flying the huge banners of Castile, standing high out of the water, with lofty "castles" forward and aft, gaudy with carving and color, the light rippling here from silken pennants and flashing there from shining cannon or huge

1 It was known later as the Invincible Armada.

poop-lanterns, moved past the southern headlands of England, watched by
half-raging, half-fearful crowds!   And how mystified and indignant must
these watching country people have been when Admiral Howard, their only
defender, calmly let the Armada sail by Plymouth, where the English fleet
lay hid in the Solent, and Captain Drake coolly insisted upon finishing a
game of bowls before he would go down to his waiting frigate.

STYLE OF SHIPS IN THE TIME OF THE ARMADA.

But these captains knew what they were about.   In those days, as now,
in fighting with sailing-vessels the advantage is usually with the one who
attacks from the windward side; for then he can manœuver his vessel,
whereas his enemy, heading toward the wind, can do so only with difficulty
if at all, and hence cannot easily take a good position or escape from a bad
one.   Howard, therefore, waited until the closely crowded squadrons of
Spain had passed beyond him up the Channel, when he issued from Plymouth harbor, bore down upon their rear from the windward, and proceeded, as one of the reports expressed it, to "pluck their feathers."
Then began some wonderful days of sea history and naval schooling.

The Spanish vessels were floating castles armed with heavy guns and crowded with soldiers armed with muskets and "harquebuses of crock,"—that is, great blunderbusses supported upon a portable rest. They kept in a close crowd, like a phalanx of old Swiss infantry, and supposed that the English would move against them in another dense raft, and that they would fight from deck to deck of grappled ships as if they were on land.

But the English knew better. They had few ships as large — the *Triumph*, 1100 tons, was the biggest — or guns as heavy as the Spaniards'. Instead of attacking in a solid mass, therefore, they spread out, hovered on the flanks, darted a ship here and there, fired as they saw opportunity, and kept their own vessels out of danger as much as possible. In the light and variable winds that prevailed, the great galleons of the Armada were almost immovable, while the English for the most part had smaller, lighter vessels, whose nimbleness and ready obedience to the helm astonished the Spanish. Standing low in the water, these would drive their shot right through the enemy's hulls, and make off before the Spaniard could depress his guns enough to do any damage in return; while the army of musketeers upon whom he had relied so strongly had little chance to do anything at all.

Thus for a week the English frigates and armed fishing-boats harassed the Armada on its way up the Channel, capturing and sinking many of the ships, while losing some of its own, of course, until at last the worried and baffled squadron managed to gain the roadstead of Calais, where the army of the Duke of Parma lay. To carry this army across and begin a campaign against London seemed now not only out of the question, but the safety of the fleet itself was a question; for a few days later, when a favorable wind arose, several fire-ships came sailing down upon them from the blockading Englishmen outside. These fire-ships — an important part of every fleet for two or three centuries — were old vessels intended to set fire to an enemy's ships. Their yard-arms were set with great iron hooks, their hulls and riggings were saturated with oil, their decks loaded with tar-barrels, and their old guns overloaded, so as to spread destruction in every direction by bursting. Then bold crews sailed these grappling monsters as near the enemy as they dared,— and it must have been a service dear to the heart of the daring,— set fire to them, lashed their helms, and got away in their boats as best they could.

To escape these dreadful things the Spaniards were obliged to up-anchor and put to sea, losing many ships and lives by fire or the wildly flying cannon-balls, or by going ashore in the effort; and then the Englishmen followed them again, like wolves after a herd of buffalo in winter. The Spaniards dared not go back down the Channel, and nothing remained to

them but the hazardous voyage around the north of Scotland—a venture
for which the towering, unwieldy galleons were ill-fitted. Storms over-
took them in the North Sea and on the Atlantic, and so many were cast

A SEA-FIGHT OF THE SEVENTEENTH CENTURY.

away on the Irish coast, where those who reached the shore were slain,
that hardly half of the proud Armada crept back to Lisbon and Cadiz.

This incident was one of the most notable in European history for two
reasons: First, historically, it no doubt saved England and her colonies
from the Inquisition, and all the other depressing and horrible burdens that
long afterward weighted the papal countries of southern Europe and their
American possessions; and, second, it reformed naval warfare not only by

confirming the value of a regularly organized national navy, but by showing that the old-fashioned, dense fleet formation, carrying soldiers to fight as they would do on land, was wrong and ineffective.

But though Spain had been humbled she was by no means crushed, and sea-fighting went on a long time before either she, the French, or the Dutch — and the last were the hardest foes — would fully admit England's claim to be sovereign of all the seas around Britain, and strike their flags whenever they met one of her "king's ships" in acknowledgment of it. England asserted that the domain of her crown covered not only the lands of England (and much of France), but also "the narrow seas"; and she defined this domain to include all the Channel waters north of Cape Finisterre and thence in a square area westward to the middle of the Atlantic. This was not an assertion: "I can beat the world in sea-fighting," but was a legal claim to rule — a declaration that her laws extended over that much sea in the same manner that it is now agreed that the laws of all nations extend to a distance of three miles from their coasts.

The whole idea of naval warfare in those days was defense of your own commerce and attack upon your enemy's; and at that time any one you met under another flag was likely to be your "enemy" if either party promised spoils worth a fight. Hence not only did privateering flourish,— often degenerating into piracy,— not only did all merchant vessels go heavily armed, but the royal ships were intended principally for convoying or guarding merchantmen. This theory, which was only a part of the generally unsettled condition of that formative period, kept up a continual state of fighting on the sea, even between peoples nominally at peace, and of course led again and again to open wars. These were almost always popular, especially among the bold sailors but poor traders of England, on account of the chances for prizes and plunder that often more than repaid the expenses and losses of the conflict; thus the war with the Dutch in 1652-54, in which William Penn was a captain, brought in more than £6,000,000 worth of captures — more than the financial cost of the war.

At this time — the first half of the sixteenth century — Holland was the leading commercial nation of the world. Not only had her merchants large interests of their own in both the East and West Indies, very extensive fisheries in northern waters, and trading stations in the African and American coasts, but a large part of the commerce of other nations was conducted in Dutch ships, including much of England itself. It was the unrighteous but determined effort to break this up by any and every means that brought on the second war with Holland, one incident of which was the capture of New Amsterdam (New York); for fleets no longer stayed close at home, acting

ATTACKING SPANISH GALLEONS OFF THE AZORES.

mainly as defenders of coasts, as in the previous century, but now cruised and fought on the high seas, as the Spanish had learned in many a hard struggle to protect their trading and treasure-ships homeward bound.

This new practice, however, had required a change in ships and their equipment. The English learned this quicker than any one else. They cut down the lofty cabins, increased the height, while reducing the weight, of masts by inventing jointed topmasts, and replaced the unwieldy lateens by an arrangement of lofty, quickly handled square sails. By the middle of the seventeenth century ocean-going ships had much the same appearance as at present,—although far more elaborately ornamented and bulging aft with stern-galleries,—the massive, high-pooped Spanish galleon surviving longest as a relic of the old type. These changes allowed the armament to be taken from the front and rear of the ship, where it had formerly been mainly placed, there being no room in the waist, and allowed it to be distributed equally up and down the ship, which now began to deliver the "broadsides" that formed such a feature in sea-gunnery before the days of turreted ironclads, and this, with the constant improvement in the range and power of the artillery, soon brought about ideas of battle formation. The early plan was to provide a large number of ships,—eighty or one hundred on each side in a single action were not uncommon,—because each was weak, and also because a great number of fighting-men was thought necessary, and then to advance from the windward in a compact mass, and

DRAWN BY ALFRED C. REDWOOD.

SPANISH AND FRENCH SHIPS OF THE LINE TAKING POSITION FOR THE BATTLE OF TRAFALGAR.

endeavor to close with the enemy and capture or destroy him by hand-to-hand promiscuous fighting. Our word *squadron* means a square, and, as applied to ships, is a survival from those antiquated methods.

But when the practice of using fire-ships became common and effective, and trimmer, more active ships superseded the cumbrous galleasses, it was seen that this close formation only exposed a fleet to destruction, and an open order had to be adopted, with a consequent change of tactics. Another lesson was, that a sea-fight was a sailor's battle, where soldiers were out of place, and that to take a great number of weak ships into action, crowded with men, was only to risk life unnecessarily. Hence, larger and more heavily armed ships, but fewer of them, appear in later engagements; and in place of a bunch of vessels, "huddled together like a flock of sheep," at which to shoot, the open order gave the gunners small and single targets.

All these changes combined to enforce the wisdom of meeting an enemy in a widely spaced line, where the strongest fighting-ships were put forward, and smaller vessels came up in the rear. Those ahead met the battle-ships at the head of the enemy's column, and the lesser ones, as they came up, were paired off against those of their own size, so that the battle became a series of equalized duels. Such was the theory of naval tactics in the seventeenth and eighteenth centuries; and so arose the term line-of-battle ship, descriptive of such national craft as are shown on the opposite page.

These fine old line-of-battle ships were large and powerful before the seventeenth century ended. Thus in the British navy when 1700 came in there were eight which had from ninety-six to one hundred and ten guns each — fifty-three others carrying more than seventy guns, and twenty-three more with more than fifty guns — all at that time regarded as fit for the line of battle, though a hundred years later nothing less than a "seventy-four" was so considered. Such were the grandly picturesque old vessels that won the day at Gibraltar, Copenhagen, and Trafalgar, and at many another spot where the whole horizon echoed to their thunderous broadsides; but of them all there now remain only a few honored hulks in harbors, or a few grand figureheads preserved in docks and museums.

Each navy, however, had a greater number of smaller, more active vessels, known as frigates, corvettes, sloops-of-war, gun-brigs, etc., which carried from twenty to forty-four guns, and were the "eyes of the fleet," as one old strategist styled them. They answered to what we should now call cruisers, and often went on duty in distant parts of the world, or in war were scouting about and supporting the main fleet. This class was especially cultivated by the United States, as soon as it began to make a regular navy, at the close of the Revolutionary War, and six frigates were built at our six navy-yards during the last years of the last century, which were intended and proved to be separately "superior to any single European frigate of the usual dimensions" in speed, manœuvering, and fighting power, in proportion to their weight of ordnance. Three of them (*Constellation, Congress,* and *Chesapeake*) mounted thirty-six guns, and three (*United States, President,* and *Constitution*) forty-four guns each — mainly 24-pounders; and all gave so good an account of themselves, as ships, that the high compliment was paid us of their being carefully imitated by foreign naval constructors.

This is not a naval history, so that I am not concerned to tell of all the glorious or inglorious work of the navies of Europe in obtaining and holding, or failing to get and keep, trade routes open and territorial possessions intact in various parts of the world. During the seventeenth and eighteenth and far into the nineteenth century, there was no time when some nations were not fighting on the sea if not on land; and much of the time *all* the maritime nations were hard at it, turning their guns to-day on the allies of yesterday, and fighting shoulder to shoulder with them the next season against some friend of the year before.

A few of the most famous battles ought to be spoken of, however, as illustrating the methods and development of naval warfare, and because we now recognize that their consequences were far-reaching.

In the wars which broke out toward the close of the eighteenth century due to Napoleon's ambition to rule the world, Great Britain found herself engaged in a struggle not only with France, but really with the whole world, for the command of the seas that washed the western coast of Europe. The only sign of friendship to England from the Baltic to Gibraltar was in the doubtful neutrality of Portugal. England had to abandon the Mediterranean, and devote herself to facing the allied powers against her outside the Gates of Hercules as best she could. In 1797 she made a beginning by crushing a fleet of Dutch ships off Camperdown (Holland), and a Spanish fleet off Cape St. Vincent; but, though both were great battles, neither had any lasting effect; and in spite of them Napoleon planned his celebrated invasion of England for the following year, supposing that by his expedition to Egypt, threatening England's East Indian possessions, he would draw away so much of the British navy that he and his allies could put an army across the English Channel unhindered. I need not say that his invasion of England never was even attempted; but for a time his fleet did hold command of the Mediterranean — a state of things to which an end was put by England's most famous naval hero, Horatio Nelson.

A long series of brilliant exploits had given Nelson fame, and the vig-

orous accounts of them he used to send home helped his great popularity.
A large part of his service had been in American waters.

In 1798 Nelson was a rear-admiral, and was sent to the Mediterranean
after the French fleet, which, having convoyed Napoleon's army to its
landing at Alexandria, was ready for new operations. It is characteristic of
the slow and almost useless methods of gaining intelligence in those days,

WHEN DECATUR WAS A MIDSHIPMAN.

that from early June to the end of July Nelson searched for this flotilla, and
was unable to get more news of it than an occasional rumor that it had been
at some place or other days or weeks before. The French knew no more
as to the movements of their pursuers, yet the fleets were twice within a
few miles of each other. This was Nelson's first independent command,
and his patience and nerves were nearly worn out by anxiety.

At last, on the first day of August, the English almost stumbled on the
French at anchor in the Bay of Aboukir, among the mouths of the Nile,

between Alexandria and Rosetta — a shallow roadstead full of shoals and
rocks, for which Nelson had neither chart nor pilot.

In the interior of this bay lay the Napoleonic squadron, under Admiral
Brueys, in such fancied security that a large part of the crews was ashore,
and some of the ships unprepared for a battle when the British appeared.
It was anchored in line of battle, however, and consisted of thirteen ships
of the line, the central one being the flagship *Orient*, having 120 guns, and
probably the largest and most complete war-ship then afloat. On each side
of her were the *Franklin* and the *Tonnant*, of 80 guns each, and none of
the others were greatly inferior.

The British had also thirteen ships, but none was the equal of the best
French, and one of them did not engage in the attack at all. Knowing
nothing of the harbor, and aware that all his ships drew much water,—
perhaps thirty feet,— Nelson had to make a long and very cautious de-
tour, throwing the lead every moment and feeling his way in. It was then
late in the afternoon, and half-past six before the *Goliath*, leading the column, got near enough to attract the French fire. Replying, but not halting,
the *Goliath*, followed closely by the *Zealous* and *Orion*, made for the head
of the line, and then with a daring unrivaled, for there was barely enough
water to float their keels, these ships slowly turned around the foremost
French vessel and dropped their anchors in the rear of the enemy's line.
The other ships, as they came up, ranged alongside the front of the French,
and the deepening twilight resounded with such a roar of broadsides as
never will be heard again.

In the darkness and smoke an English seventy-four, the *Bellerophon*,
had engaged the monstrous *Orient*, and in a short time had been crushed :
all her masts were swept out of her, two hundred of her people were
killed and wounded, and she drifted out of action. But nearly the same fate
had by that time overtaken the French *Guerrière*, for the *Theseus* had coolly
placed herself where she could rake the anchored ship and tear her to pieces.
The moment the *Bellerophon* drifted off, however, her place was taken by
two newly arrived frigates, and the *Orient* presently found herself the target
of three ships which slowly but surely were cutting her to pieces in spite of
her tremendous resistance. Her admiral had been killed on her deck, where
half her officers and men lay dead or wounded, when it was suddenly seen
that she was on fire, and the whole battle was instinctively suspended to
watch the magnificent spectacle, save where some still poured in shot and
shell to prevent the French crew from extinguishing the flames.

Powerless either to save their ship or launch their boats, the remnant
of the *Orient's* crew could only fling themselves into the water and trust

to the mingled boats of friends and foes to pick them up. The ships nearest slipped their cables, and tried to edge away out of danger as the flames enveloped the towering masts, burning with amazing fierceness in the tarred rigging and lighting up the desert for miles inland, while the hull became a furnace. Suddenly, at a quarter before ten, a volcano-like explosion tore the glowing old battle-ship asunder, a torrent of burning fragments was

THE "THESEUS" ATTACKING THE "GUERRIÈRE."

hurled aloft,— with how many dead heroes, no one knows,— and double darkness closed over the appalling scene. Then the black waves were lighted anew by the flash of cannon and musketry, and the battle went on until daylight before the last of the French vessels had been conquered, while two of them had managed to steal away. Of the other eleven one had been burned and sunk, three had gone ashore, where one burned, and the remainder had been crushed into surrendering. The English did not lose a single vessel, for even the dismantled *Bellerophon* could float, and their loss in men was far less than that of the French.

Historians tell us that this victory was the grandest naval success on record. Nelson himself said that victory was too weak a term — it was a catastrophe. It put an end at once to Napoleon's campaign in Egypt, and to all his designs against India. It gave the command of the Mediterranean to England, emboldened Turkey and Russia to recover the Ionian Islands, gave Naples a chance to assert herself, and aroused Austria and Russia to resist by armies Napoleon's aggressions, so that from this battle dates his downfall. Its influence soon reached the United States, and caused it to break through its neutrality and begin upon the sea that naval war with France of which we hear very little nowadays, but which gave to our own naval record such glorious incidents as Truxton's battles in the *Constellation* with *L'Insurgente* and *La Vengeance*, and Captain Little's capture, in the corvette *Boston*, of the French sloop-of-war *Le Berceau*.

Nelson remained in the Mediterranean for some years, by no means idle, and then did service of extraordinary value elsewhere, as at the battle of Copenhagen, which in a single remarkable conflict put an end to a northern conspiracy against England, and saved her a vast deal of trouble; but his final service was the most momentous of all, at any rate for the fortunes of Great Britain alone, and this was the winning of the battle of Trafalgar.

In 1805 Napoleon had prepared for another grand invasion of England, and with great skill had gathered a fleet of allied French and Spanish vessels, which was to protect and coöperate with the strong army he proposed to land along the Kentish shores. This fleet was commanded by Admiral Villeneuve, and assembled at Cadiz, where, in October, 1805, it was being watched by an English fleet, commanded by Nelson and Collingwood, consisting of thirty-three ships of the line; twenty-seven of these were present when, on the morning of the 21st, the allies, twenty-nine battle-ships strong, came sailing out, hoping to avoid battle if possible. This, Nelson was resolved, should not happen; and dividing his forces into two columns, he made at them in such a way as to strike their line (then off Cape Trafalgar) in the middle of its crescent. The wind was very light, and an hour or more elapsed before even the heads of the line struck the enemy, so that there was plenty of time to make every preparation, and there was constant instruction by signaling from Nelson's flagship *Victory*. Then at the last moment, when the first gun was ready to be fired, there rose upon the signal halyards of the *Victory* the message that, received with ringing cheers, has been an inspiration to patriots the world around ever since —

ENGLAND EXPECTS EVERY MAN WILL DO HIS DUTY.

A few moments later Collingwood in the *Royal Sovereign*, and Nelson in the *Victory*, were in the thick of the foreign fleet, which awaited them in disorderly array, but closed about these two, bent upon destroying them if possible before any others could come up. The fury of the duels that ensued, where ships were mixed in disorder, and sometimes three or four against one, passes adequate description. None, perhaps, fared worse than

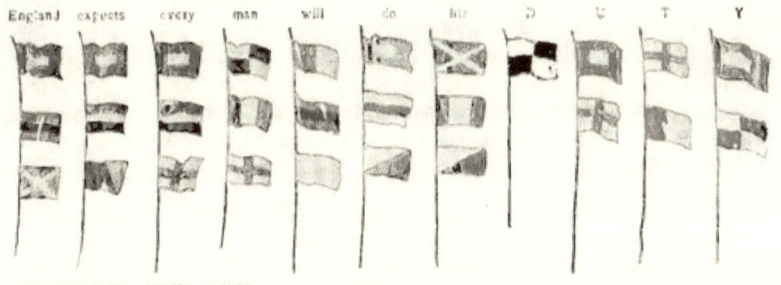

England   expects   every   man   will   do   his   D   U   T   Y

DRAWN FROM THE MODEL IN THE GREENWICH MUSEUM   NELSON'S SIGNAL.

the *Belle Isle*, a large English two-decker that was the first to reach
the scene after the *Royal Sovereign*, and to draw off some of the fire that
threatened to pulverize Collingwood's ship.

The wreckage and suffering on other ships were almost as great. The
very first broadside of the *Royal Sovereign*, taking the *Santa Ana*, struck
down 400 out of the 1000 persons aboard; and the *Sovereign* herself soon
lost every mast. The *Santissima Trinidada*, a Spanish four-decker, and
the largest ship then afloat, was reduced to a wreck, and a dozen others
lost a part or all of their masts. As for the *Victory*, she was always in the
thick of it, receiving at one time the concentrated fire of seven hostile battle-ships, yet was not too much disabled to be manœuvered. Her captain's
aim was to engage directly with the French flagship *Bucentaure*, but she
was closely attended by three other large ships, and difficult to reach.
Nevertheless, the *Victory* finally got across her stern, and from a few yards
distance poured in a broadside which, sweeping the whole length of her
interior, dismounted twenty guns, and killed and wounded 400 men. As
she passed on, returning the fire of the other vessels near by, she was
closely followed by the *Temeraire*, the second English ship, which had
already become almost unmanageable; and a lifting of the smoke showed
her smashing a little French frigate, the *Redoubtable*, which, by and by, was
captured after almost every man had been killed, and she was in a sinking
condition. The astonishing resistance of this little vessel, and the damage
she did by soldiers with muskets crowded in her tops and firing down upon
the decks of the English ships, form one of the most noteworthy incidents
of naval history; and it is not too much to say that she inflicted upon Great
Britain as great harm as all the rest of the allies put together, for it was a
musket-ball from the mizzentop of the *Redoubtable* that struck down, early
in the action, the great Nelson himself. He seemed to have had a feeling,
even before leaving England, that he would not survive this campaign, and

knew his wound was mortal the instant it was received. He was carried below, and remained alive and conscious about three hours, eagerly listening to reports of the progress of the fight, and rejoicing at last in a knowledge of victory. His last words, murmured again and again, with his failing breath, seemed an answer to his signaled injunction, for they were: "*Thank God I have done my duty.*"

Other men [writes Captain Mahan] have died in the hour of victory, but to no other has victory so singular and so signal stamped the fulfilment and completion of a great life's work. " Finis coronat opus " has of no man been more true than of Nelson. Results momentous and stupendous were to flow from the annihilation of all sea power except that of Great Britain, which was Nelson's great achievement; but his part was done when Trafalgar was fought, and his death in the moment of completed success has obtained for that superb victory an immortality of fame which even its own grandeur could scarcely have insured.

No such fleet actions as this ever occurred in North American waters in the time of the "old navy," though there was plenty of cruising and fighting up and down the coast and in the West Indies. The United States had made its new flag respected before the end of the eighteenth century, but it was done mainly in European waters, where that marvelous captain, Paul Jones, had been defying enemies to the point of rashness.

Paul Jones was the first man to hoist our national ensign (the rattlesnake flag) on an American ship, and again the first to hoist the stars and stripes, and was the ranking officer of the continental navy. He records that "in the Revolution he had twenty-three battles and solemn rencounters by sea: made seven descents in Britain and her colonies; took of her navy two ships of equal and two of far superior force," and so on. It is true that he alone of his day steadfastly refused to acknowledge England's supremacy of the seas; that the flag of the United States alone was never struck to Great Britain except under force of honorable combat; and that on the ships commanded by Paul Jones it was never struck at all!

Every Yankee school-boy knows of the terrible fight of the crazy old sloop-of-war *Bon Homme Richard* against the *Serapis*, a new English 50-gun frigate in the North Sea, in which a sinking and burning and shot-riddled vessel, able after the first broadside to bring only three or four small guns into practice, conquered and captured her twice-greater antagonist. It is not a story one can tell in a few words, but it was a deed that is regarded in naval annals as among the most extraordinary in the history of the world, and it won for the new republic a credit in Europe that was of vast benefit to it and all its wandering citizens.

Great Britain, though humiliated, had not been seriously hurt by the loss of two or three ships out of her six hundred, and she still tried to enforce against the rising naval power on the west side of the Atlantic the subservience which she received along its eastern shores. It took the form of asserting her right to stop and board any American vessel, governmental or private, and seize and impress into her own service any British subject found serving in the crew. This always met with protest and resistance, and at last became so galling that in 1812 the United States declared war against Great Britain's might rather than continue to submit to it.

This might gradually overcame us, and British fleets sailed up and down our coasts unhindered, but not until the enemy had been surprised by many harder knocks than they anticipated, and had learned one thing for certain,— that while man for man the Yankees were equally good seamen and fighters, they were better ship-builders, and could teach lessons in that art which their enemies were not above learning; and finally we won by sheer force of victories at sea.

BARON NELSON OF THE NILE.

I have already spoken of the six frigates which were used in that war, as admittedly the best of their kind in the world. Except the unlucky *Chesapeake*, which was rashly carried unprepared into the fatal action against the *Shannon*, where Lawrence lost his life, but won undying fame in the memory of his countrymen by his "Don't give up the ship," all did glorious work. Thus, the *United States* under Decatur reduced to a wreck off Madeira, and brought as a prize to New York, the British 44-gun frigate *Macedonian* in October, 1812, itself remaining almost uninjured,— a victory due to superior seamanship and gunnery.

The same skill, using a ship of superior sailing power, accounted largely for the splendid victory of the United States sloop-of-war *Wasp* (18 guns), a week earlier, near Bermuda, in an encounter with the British sloop *Frolic* (19 guns), where in three quarters of an hour the *Frolic* was totally dismasted and reduced to a rolling wreck, with ninety killed or wounded out of a crew of one hundred and ten, while the *Wasp's* loss was only ten. A British seventy-four then came up and captured both the victor and her prize; but eighteen months later a second *Wasp*, by reason of her better gunnery, cut to pieces at different times two other ships with compara-

9

THE "FROLIC" REDUCED TO A WRECK BY THE FIRST "WASP" (1812).

tively small injury to herself. Nor could the *President* have given so good
an account of herself in her unfortunate encounter with the *Belvidera*, and
again when chased and finally captured by the squadron led by the *Endy-
mion*, had not her sailing qualities and gunnery been of so high an order —
qualities which also distinguished the American fleets on Lake Erie and
Lake Champlain.

But the honors of that brilliant naval war belonged chiefly, after all, to
the *Constitution* — "Old Ironsides," as the people loved to call her, — which
is enshrined in the history and hearts of the United States as Nelson's
Victory is in those of Great Britain.

The *Constitution* was the finest, perhaps, of the United States frigates,
and a favorite ship with commanders, yet her fame began with her success
in running away, Broke's British squadron chasing her three nights and
two days, only to lose her after all. The winds were so light that she sent
out her boats to help the sails urge her forward. It was only a few days
after that (August 19, 1812) that Commodore Isaac Hull, cruising in search
of the British vessel *Guerrière* (the same that had been captured from the
French in the battle of the Nile, and again dismasted at Trafalgar), over-
hauled her off the coast of Newfoundland. The London newspapers had
not only been sneering at the *Constitution* as "a bundle of pine boards sailing under a bit of striped bunting," but Captain Dacres had sent a boastful

challenge to Hull to meet him and see what would happen. The vessels, though nominally of different rate, were actually in close equality, and both crews were eager for a fair fight. It was already well along in the afternoon, and the sea was rough, but Hull would not reply to the enemy's fire until he was within pistol-shot, then his broadside opened.

" Fifteen minutes after the contest began," to quote Lossing's lively account, " the mizzen-mast of the *Guerrière* was shot away, her mainyard was in slings, and her hull, spars, sails, and rigging were torn to pieces. By a skilful movement, the *Constitution* now fell foul of her foe, her bowsprit running into the larboard quarter of her antagonist. The cabin of the *Constitution* was set on fire by the explosion of the forward guns of the *Guerrière*, but the flames were soon extinguished. Both parties attempted to board, while the roar of the great guns was terrific. The sea was rolling heavily, and would not permit a safe passage from one vessel to the other. At length the *Constitution* became disentangled, and shot ahead of the *Guerrière*, when the main-mast of the latter, shattered into weakness, fell into the sea. The *Guerrière*, shivered and shorn, rolled like a log in the trough of the billows. Hull sent his compliments to Captain Dacres, and inquired whether he had struck his flag. Dacres, who was a 'jolly tar,' looking up and down at the stumps of his masts, coolly and dryly replied : ' Well, I don't know. Our mizzen-mast is gone, our mainmast is gone,— upon the whole you may say we *have* struck our flag.' "

Too completely wrecked to be of any further use, the historic old ship was set on fire and blown up, and so ended her pride and her story. Hull lost only fourteen men killed and wounded, while the British lost seventy, dead, and all the survivors prisoners. This calamity, on the heels of similar successes elsewhere for the "bit of striped bunting," spread consternation throughout Great Britain not only, but in the other European monarchies, for it presaged the rise of a new power to be reckoned with, where novel and superior instruments and methods of warfare opposed uncalculated forces to the old régime.

This conviction was enforced upon Europe anew only four months later by the *Constitution* overtaking and crushing in West Indian waters the 38-gun frigate *Java*, which also was burned to the water's edge, because the wreck was not worth saving; and again the British loss was many times greater than the American. Captain William Bainbridge, who had distinguished himself in the Mediterranean, was her commander.

Various successes marked her career for the next two years, until, under the command of Captain Charles Stewart, she had her memorable adventure off Madeira, in which she engaged with the two British ships *Cyane*, thirty-six guns, and *Levant*, eighteen guns, and captured both, with a loss of only three men killed and twelve wounded. Stewart set sail with his prizes and prisoners for Porto Praya, whence he purposed sending his prisoners to New York in a captured merchantman. Reaching there on

THE "CONSTITUTION" CHASED BY CAPTAIN BROKE'S SQUADRON

The ports on the upper deck aft were roughly cut to meet the emergency.   The sailors in the rigging
threw water from buckets upon the sails to make them hold better the faint breeze, and
below hose pipe was used to the same purpose.   During the three days'
chase boats were sent out to tow, and kedge-anchors
were used to warp the ship forward.

March 10th, he was next day busy at these arrangements, when the topsails
of several men-of-war were seen entering the harbor through the prevailing
fog.   Having no trust that, if these were British, their commanders would

respect the courtesies of a weak neutral port, Stewart felt that his only chance was to try to run away in the fog, and made immediate preparations to do so, sending word to the *Levant* and *Cyane* to follow. Being discovered by the strangers — three large British frigates — at the outlet of the harbor, their escape immediately became a question of seamanship and sailing. Here the Americans showed their superiority, and effectually dodging both the ships and the cannon-balls of the pursuers, the *Levant* got back under the protection of the guns of the fort at Porto Praya, while the *Constitution* and *Levant* fairly outsailed the frigates and escaped.

In 1830 brave Old Ironsides was condemned as worn out, and ordered to be sold. But, as a similar sad fate overtaking the "Fighting *Temeraire*" had been made the occasion of an immortal painting by Turner, and so, perhaps, had caused Nelson's still more famous battle-ship *Victory* to be preserved in the harbor of Portsmouth as a shrine of naval inspiration, so the obloquy that menaced the *Constitution* now fired the heart of a young poet to write a passionate appeal to patriotism. Who does not know Dr. Holmes's ringing stanzas? —

Oh, better that her shattered hulk
    Should sink beneath the wave;
Her thunders shook the mighty deep,
    And there should be her grave.
Nail to the mast her holy flag,
    Set every threadbare sail,
And give her to the God of Storms,
    The lightning and the gale!

HOMEWARD BOUND.

The country caught the spirit, and such a cry of protest went up that the vandalism was stayed, and Old Ironsides was again repaired — hardly anything but her ornaments was now left of the original structure — and took several cruises, one of which was in carrying wheat to famine-stricken Ireland. Later she was used as a school-ship, but finally became worthless even for that, and in 1895 the question arose whether she should be broken up at the Brooklyn navy-yard or towed around to Portsmouth, New Hampshire, and there laid up in a line with the *Macedonian* and a few other ancient hulks that were rotting quietly away in honorable age, and have now wholly disappeared. Sentiment dictated the latter course, and, with a crew aboard, prepared to take to their boats at a moment's notice, the leaking and crazy old warrior, stately even yet, and sadly saluted by every fort and vessel she passed, crept around to her last berth at Kittery Point. She is the last and the most glorious representative of the "old navy."

H. J. R. DOVER 1890

TYPES OF BATTLE-SHIPS — 1890 AND 1800.

CHAPTER VI

(Continued)

WAR-SHIPS AND NAVAL BATTLES

PART II — THE PRESENT ERA OF STEAM AND STEEL

HE introduction of steam made little difference in naval affairs at first, so far as either strategy or tactics are concerned, although it changed the conditions of naval action in two principal ways and in many minor ones. Ships could now, like the early galleys, be placed in any position the commander pleased, and, unlike galleys, this effort could be sustained a long time, for engines do not tire out like human arms. On the other hand, ships propelled by steam needed to return to port at frequent intervals to obtain coal, and naval powers found it necessary to provide, either by possession or treaty, safe coaling-stations in various parts of the world for the use of their cruising fleets.

The first steam war-ships were naturally fitted with side paddle-wheels: but as soon as the screw-propeller came into use the navy was quick to adopt it. "By its use the whole motive power could be protected by being placed below the water-line. It interfered much less than the paddle with the efficiency and handiness of the vessel under sail alone, and it enabled ships to be kept generally under sail. Great importance was attached to this, as the handling of a ship under sail was justly thought an invaluable means of training both officers and men in ready resource, prompt action, and self-reliance." For this reason masts and sails were retained long after they were admitted to be detrimental to the fighting qualities of battle-ships. Naval reformers had to wait until the last generation of "old salts," trained on "blue water," had died off, and their scornful sneers at "tea-kettle" seamanship had been silenced in the only way possible, before they could persuade governments to build or men to serve in the new style

THE "KEARSARGE" GETTING INTO POSITION TO RAKE THE "ALABAMA"
AT THE CLOSE OF THE COMBAT.

of vessels. In truth, the transition from the fighting machinery and methods that prevailed until, say, the bombardment of Acre, in 1840, to those
that decided the inferiority of China in her struggles with Japan at the
Yalu and elsewhere, was rapid enough to make even a sea-dog dizzy.

Excellent types of the war-steamers, intermediate between the old two-
and three-deckers and the sailless "ironclads" that followed, were those
two actors in that most glorious sea-fight of the American Civil War—
the *Kearsarge* and *Alabama*.

In this great fight, which took place a few miles off the harbor of Cher-
bourg, France, one beautiful summer Sunday (June 19th) in 1864, much the
same tactics prevailed as in any one of the earlier ocean duels. As the
Alabama came on she began firing the two-hundred-pound pivot-rifle forward, which was her main gun, while the *Kearsarge* was yet a mile away.
The latter waited a little before replying, but only a few moments elapsed
before both were near enough and hard at it, each doing its best to get a
position ahead of its antagonist for raking,—a disadvantage which the
other steadily avoided; and this caused them to follow one another about in
advancing circles, of which seven were described before the end came.

We have a story of the battle as seen from the deck of the *Kearsarge*,
written by her surgeon, who had little to do except observe the conflict.

The *Kearsarge* gunners [he tells us] had been cautioned against firing without direct aim,
and had been advised to point the heavy guns below rather than above the water-line, and to
clear the deck of the enemy with the lighter ones. Though subjected to an incessant storm
of shot and shell, they kept their stations and obeyed instructions.

The effect upon the enemy was readily perceived, and nothing could restrain the enthusiasm of our men. Cheer succeeded cheer; caps were thrown in the air or overboard; jackets were discarded; sanguine of victory, the men were shouting as each projectile took effect: "That is a good one!" "Down, boys!" "Give her another like the last!" "Now we have her!" and so on, cheering and shouting to the end.

After exposure to an uninterrupted cannonade for eighteen minutes without casualties, a sixty-eight-pounder Blakely shell passed through the starboard bulwarks below the main rigging, exploded upon the quarterdeck, and wounded three of the crew of the after pivot-gun. With these exceptions, not an officer or man received serious injury. The three unfortunates were speedily taken below, and so quietly was the act done, that at the termination of the fight a large number of the men were unaware that any of their comrades were wounded. Two shots entered the ports occupied by the thirty-twos, where several men were stationed, one taking effect in the hammock-netting, the other going through the opposite port, yet none were hit. A shell exploded in the hammock-netting and set the ship on fire; the alarm calling for fire-quarters was sounded, and men detailed for such an emergency put out the fire, while the rest stayed at the guns.

The *Kearsarge* concentrated her fire and poured in the eleven-inch shells with deadly effect. One penetrated the coal-bunker of the *Alabama*, and a dense cloud of coal-dust arose. Others struck near the water-line between the main and mizzen masts, exploded within board, or passing through burst beyond. Crippled and torn, the *Alabama* moved less quickly and began to settle by the stern, yet did not slacken her fire, but returned successive broadsides without disastrous result to us.

Captain Semmes witnessed the havoc made by the shells, especially by those of our after pivot-gun, and offered a reward for its silence. Soon his battery was turned upon this particular offending gun for the purpose of silencing it. It was in vain, for the work of destruction went on. We had completed the seventh rotation on the circular track and begun the eighth; the *Alabama*, now settling, sought to escape by setting all available sail (fore-trysail and two jibs), left the circle, amid a shower of shot and shell, and headed for the French waters; but to no purpose. In winding the *Alabama* presented the port battery with only two guns bearing, and showed gaping sides through which the water washed. The *Kearsarge* pursued, keeping on a line nearer the shore, and with a few well-directed shots hastened the sinking condition. Then the *Alabama* was at our mercy. Thus ended the fight after one hour and two minutes.

One incident of this battle much talked of at the time, and given as an excuse for their defeat by the Confederates (though without good reason), was the fact that the waist of the *Kearsarge*, opposite the engines, was protected by anchor-chains, hung in close festoons on the outside of the ship, and kept in place and concealed by a boxing of thin boards. This, however, was not the first attempt at protecting ships by armor, which had now become necessary to meet successfully the better guns and projectiles that year by year were increased in penetrative power. New powders and explosives were constantly being invented also, each more effective than the preceding; and as these not only used in guns but applied to the filling of shells, these bursting missiles for a time almost displaced solid shot.

Along with this the discovery and perfection of the Bessemer and other processes of making steel, and methods of adapting rifling to great cannon, produced a rapid and varied increase in size and an improvement in quality in the guns supplied to ships as well as in those used upon shore.

Against these new weapons the old "wooden walls" were of no avail. Oak and teak, however sound and thick, failed to turn aside the conical projectiles as they had the old round shot and shell. The ponderous missiles would crash clear through, smashing everything in their path, and sending showers of death-dealing splinters right and left. The navy had to protect itself by a revival of the armor with which knights of the middle ages guarded against arrows and javelins and sword-points. By and by, when

THE UNITED STATES FRIGATE "MERRI-
MAC" BEFORE AND AFTER CONVER-
SION INTO AN IRONCLAD.
Compare with illustration on page 139.

guns and bullets came, the knights thickened their armor in an attempt to resist these new missiles, until at last it reached a weight too great to be carried, and the whole cumbrous panoply had to be laid aside, and knightly tactics altogether changed. Many persons believe that this history will be repeated in the case of the sea-warriors of the world, which, within the memory of many a grizzled admiral, have changed from buoyant and beautiful ships to grim and shapeless fortresses afloat.

The Americans, fearless of sea-traditions, were the first to propose armor for ships, but the French first practically applied it, building several "floating batteries," covered with iron 4¾ inches thick, in 1855. The English copied them, in somewhat more ship-shape form; and then the French began boldly to sheathe some of their frigates with iron plates and call them "ironclads." By this time iron hulls had begun to be used commonly in the British merchant service, but of course the men-of-war's men, the slowest class of persons on earth to accept any change, insisted that iron would by no means do for war-ships. Nevertheless a few progressive spirits persuaded their high-mightinesses, the Lords of the Admiralty, to try an experiment in building one, and, in 1860, the first iron war-ship was

launched and named *Warrior*, while all the old salts wagged their heads
and predicted the end of " Britannia rules the waves," until there was n't
a really *jolly* tar to be found from Penolar Point to Pentland Firth.   To a
certain extent these hardy old growlers were right, though their idea of a
remedy was wrong.   It proved a failure to build old-style battle-ships of iron
or even of steel, or to coat them all over with armor, even when greatly
thickened.   Not only were they slow and somewhat unmanageable, but by
the time one of them had been built with thicker walls than its latest rival,
somebody had invented artillery whose projectiles would penetrate it.
Ships that are " ship-shape," that is, possess masts and sails, but are constructed wholly of iron or steel, and more or less heavily armored, have survived, and will always be a part of the world's navies, no doubt, but their
uses will be subsidiary to heavy fighting; and with the disappearance of the
wooden sailing line-of-battle ship in the Crimean war and of the iron war-
steamer a quarter of a century later, all traditions of the " old navy " were
ended — traditions that went back to the days of Drake.

But who could have foreseen that this swift and momentous upsetting
should come about, not through the efforts of the great sea powers of Europe, — the giants who had been struggling for the control of the ocean for
three hundred years, — but from the brain and purse of landsmen in a country of the New World not taken into account as a naval power at all.

You need not be told that it was Ericsson's invention and Henry Grinnell's building and Lieutenant Worden's courageous fighting of the little
Monitor in Hampton Roads,
on that fair March Sunday
in 1862, that brought about
this change.   When her turret — the "cheese-box on a
raft " — successfully withstood
the assault of that heavily
armed floating battery, the
Merrimac (or *Virginia*), all
the war-ships of the world felt
themselves beaten, too, and
wise seamen saw that they
must prepare to face a new foe.

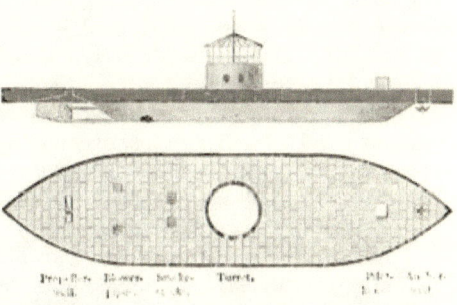

SIDE ELEVATION AND DECK-PLAN OF THE
" MONITOR. "

At once all maritime governments began to build fighting-vessels which
were castles of steel afloat, and smaller ships for various services that more
resembled a Nootka war-canoe in outline than one of the frigates that used
to do their work.   So shapeless were they that a new term had to be used,

and we began to call them *cruisers*. All war-ships, in fact, are now classified by their work, not by their shape or size or rig.

First, fewest, and heaviest are the harbor-defense vessels — monitors and massively walled floating batteries, intended to remain in harbors, or close to the coast, as movable forts.

Second, battle-ships — the strongest, most thickly armored, heavily armed style of ships that can be made, and still be able to go to sea; but these are not expected to leave their home ports for a long time, nor to go to any great distance unless compelled to do so in actual war.

Third, cruisers. These take the place of the old-fashioned lesser fighting-ships, the seventy-fours, frigates, corvettes, and sloops, and vary greatly in size, model, speed, and power of armament.

Fourth, small, swift, strongly armed but lightly armored, torpedo-boat chasers, small gunboats for use in rivers and shallow coastal waters, despatch-boats, dynamite-cruisers, such as our American *Vesuvius*, tow-boats, and similar minor craft — the run-abouts of the naval service.

Fifth, torpedo-boats.

The material of all these is steel. Wood is no longer permitted even in the fittings of their cabins, because wood will splinter and burn.

The great hull of a modern battle-ship, as described by Lieutenant S. A. Staunton, U. S. N., which supports and carries the vast weights of machinery, guns, and armor, aggregating perhaps more than ten thousand tons, is built of plates of rolled steel, varying from $1\frac{3}{4}$ inches thick at the keel to $\frac{3}{4}$ inch at the water-line. These are closely jointed and fitted, and bound together with straps, angle-irons, and brackets, so as to make a strong unyielding structure braced in all directions. Then, through the central part of the ship, at least, vertical plates are erected upon the frame and outside plating, which bear a second or inner bottom, thus forming the "double bottom" as high as the water-line, having the space between the inner and outer sheathing separated into a multitude of small water-tight cells, so that an injury to the outside hull would not cause the vessel to leak unless the inner bottom were also punctured.

Throughout the whole length of the vessel, reaching from side to side and from the keel to the main deck, are many steel bulkheads, sufficiently strong to resist the pressure of the water, and communicating only by water-tight doors, so that even were an accident, such as a collision or running upon a rock, or an enemy's shell, to open a hole through both bottoms, the ship would still float, because the inflowing water would be confined to a single compartment, leaving the rest of the ship dry and buoyant. Nothing less than the blow of a ram, smashing through everything and throwing several compartments into one, would be likely to sink such a ship, and this is one reason why ramming has again become prominent in naval tactics.

But while safety from sinking is thus reasonably assured, this is more a precaution of seaworthiness against the accidents of storms than toward injuries receivable in battle. Passenger and freight steamers now

THE FIRST SEA-FIGHT OF MODERN WAR-SHIPS.

The Peruvian turret-ship "Huascar" between the fire of the Chilean ironclads "Almirante Cochrane" and "Blanco Encalada," October 7, 1879.

have the double bottoms and water-tight compartments, and the best of these have arrangements for mounting light but powerful guns upon their decks, so that they may be utilized by the government in a war emergency as light cruisers, as armed transports, as swift scouts, or in other highly important ways; they will then be coated with a light protective armor, but will not be expected to engage in a contest with a real fighting-vessel.

The idea of armor-plate is, as has been said, scarcely half a century old, and the moment it was put on (amid the jeers of the old line-of-battle tars, who thought they had done all that the dignity of the profession permitted when they arranged their rolled-up hammocks along the bulwarks to catch musket-balls, and spread nettings to prevent somewhat the flight of splinters) ingenious men began to improve their powder and strengthen their guns to overcome the new defenses. To meet these improvements armor has been increased and perfected, until now war-vessels are no longer "ships" in any proper sense of the word, but floating fortresses of steel, the names of whose defensive parts, even, have been borrowed from land fortifications, such as *turret* and *barbette*.

A limit to this defensive strength is marked in two directions. First, by the size it is possible to make a vessel, and still keep her seaworthy

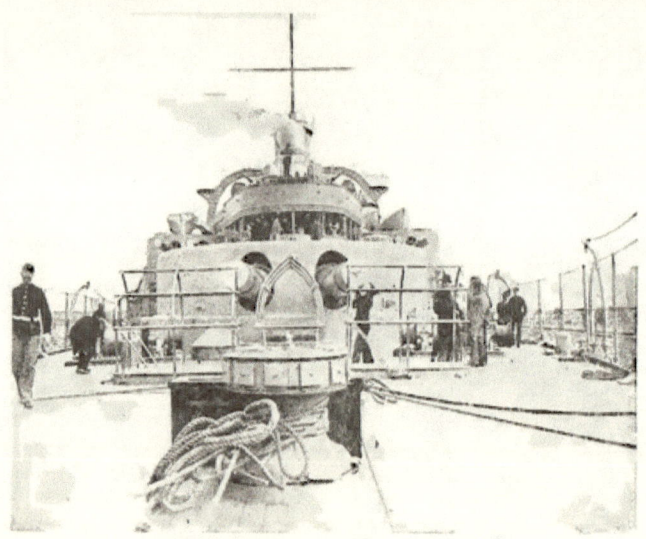

THE UNITED STATES BATTLE-SHIP "MASSACHUSETTS."

and manageable; and, second, by the weight of armor such a vessel can carry, in addition to the weight of the framework, machinery, guns, and other things necessary. These limits seemed to be reached some time ago in some of the monstrous battle-ships built in Europe, and when it was found that even while they were in construction rifled guns had been invented that would drive their projectiles through the thickest wall of wrought-iron or steel that these or any other vessels could carry, naval constructors began to despair of keeping ahead of the gun-makers, and there was even talk of abandoning armor altogether, and fighting battles out with bared breasts as we used to do.

The percentage of weight which may be allotted to armor in the design of a ship limits the area which can be wholly protected, but often permits the partial protection of other areas of less importance to her vitality and destructive force. Motive power, steering-gear, and magazines stand first upon the list of those features demanding complete protection. . . . The heavy shells from an enemy's guns may do many other forms of injury besides sinking a vessel and disabling her crew. They may strike and disable her engines, or pierce her boilers, causing disastrous explosions. They may injure her steering-gear, destroy the mechanism which controls her turrets and guns, or injure the guns themselves and their carriages. In every feature of offense which renders her a formidable and dangerous foe — her speed, her mobility, the fire of her guns — a man-of-war is dangerously vulnerable unless she be protected by armor, unless the enemy's shot be rejected by plates which it cannot penetrate.

Then came an invention that put a new face upon the matter, — the surface-hardening of plates, composed of a mixture of nickel with steel, — which, from one of its perfectors, is known as "Harveyizing" it. Other processes also are known. This gave to the surface of the metal such a flinty hardness that the heaviest and most highly tempered steel projectiles would almost invariably break to pieces when they struck it — the same projectiles that were able to punch a hole clear through a target-plate of ordinary wrought-steel twenty-two inches thick!

Plates thus surface-hardened are now made in Europe, and as well, if not better, in the United States, where we have learned and taught the rest of the world how to make them by rolling — a much better, as well as cheaper, process than the former method of hammering them into shape.

It was found that with these hard-surfaced plates much less thickness was required to contend successfully with the great guns opposed to them than had been the case before; and the great saving of weight enabled a much larger extent of armor to be borne upon a ship than was formerly possible, so arranged as to protect all her hull and vital parts.

Thus, in a typical modern battle-ship, say 360 feet long, 72 feet broad, and drawing 24 feet of water, having an armor of surface-hardened nickel-steel, this armor is thus disposed: amidships, and a quarter of her length behind the point of the prow, is built up a semicircular "barbette," or wall, of the thickest armor, behind which is a "turret," moving to the right or left through an arc equal to half the horizon, no higher than necessary to cover and work the guns, and having its motor mechanism fully protected by the barbette. This is the forward turret — a swinging fort, carrying with it, as it turns, two of the heaviest guns in the ship.

Half-way from the center to the stern stands the after turret and its

THE UNITED STATES BATTLE-SHIP "INDIANA."

barbette, similarly built of the strongest armor,— ten to twelve inches thick,— and sweeping with its guns half the horizon.

From a point just in front of the forward barbette two walls of the heaviest possible armor, reaching vertically from four and a half feet below the water-line (loaded) to three feet above it, extend diagonally backward to the sides of the ship, then continue along its side in a "belt" to points opposite the after barbette, where they bend inward as before and meet just aft of the after barbette ; but hereafter the increased efficiency of armor, by further reducing its weight, will probably enable the armor-belts to be carried to the extreme ends of the ship, which otherwise can be so seriously damaged by an enemy as to interfere with the speed and control of a ship in action, even if it does not disable her.

But while these upright walls will resist a direct shot, it is equally necessary to guard against a plunging fire, and therefore the space between the turrets, at least, must be roofed over with a steel deck, two or three inches thick, to deflect shot that come just over the top of the armor-belt.

In addition to this, on each side of the vessel are erected one or two smaller turrets, carrying somewhat smaller guns than those of the forward and after turrets, and also protected by heavy barbettes which reach down to the armor-belt and thoroughly protect the turning mechanism, passage of ammunition, etc. These various upper parts are connected by defenses which may not resist the largest shells, but are safe against smaller shot.

Now, what is the armament of this fortress which thus protects all the motive power and interior machinery of the ship, by which she can be made so terrible an engine of combative force? Well, it is as different from the bronze "long-toms" and carronades of the old three-deckers, or even from ten-inch smooth-bore "Dahlgrens" of the days of our Civil War, as is the ship itself from old-time models. In place of broadside batteries of forty or fifty cannon hidden in clouds of smoke, there are now six or eight big rifles, from whose muzzles wreaths of thin gas only drift to leeward : and, more striking still, in contrast, a ship is no longer comparatively helpless when headed or turned sternward to an enemy,— when the "raking," formerly so justly dreaded, would be received,—but is rather more able to do damage in that position than by a "broadside."

The guns themselves are marvels of structure and power. All of those used in the United States navy are made by the government in the gun-shops at the Washington navy-yard, and are "built up." The methods and tools required for this are the invention of Americans, as well as the complicated arrangements for closing the breech, and the carriages and mechanism for overcoming the tremendous recoil and handling the ponder-

10

ous ammunition; the latter, often weighing hundreds of pounds, is handed up to the gunners from the magazines below by hoists worked by electricity.

The history of the development of heavy ordnance, especially that applied to naval uses, is one of the most interesting chapters in mechanics; and a surprising number of ways of making a ship's cannon have been tried and rejected. Out of this two things seem now to be settled: namely, that a gun composed of steel in separate parts welded together is best, and that the best missile to shoot from it is a conical shell, very hard and heavy, yet containing an explosive small in quantity but exceedingly powerful.

Such guns are built up of a tube or "core" of steel of the required size, upon which is shrunk a jacket, covering the rear, or breech half of the core, outside of which are shrunk on several broad hoops. The cutting out of the bore to exactly the proper caliber and the plowing of the spiral riflings

THE UNITED STATES CRUISER "BROOKLYN" (STERN VIEW).

put the gun in readiness for its breech-closing and other attachments. This process requires several months, involves large capital and powerful machinery, and good results imply the very highest workmanship.

Such are the guns of modern men-of-war; and a first-class battle-ship carries four twelve- or thirteen-inch rifles (that is, having a bore twelve or thirteen inches in diameter), several eight- or ten-inch rifles, and many

smaller guns arranged to be fired with extraordinary speed, and hence called
"rapid-fire" guns; while her upper works and "military tops" fairly bristle
with fierce little six-, four-, and one-pounders,—revolving magazine rifles,
capable of discharging rifle-balls as fast as a man can turn the crank.

ON BOARD A BATTLE-SHIP GOING INTO ACTION.   WORKING THE RAPID-FIRE GUNS.

To give some idea of the size and power of one of the 13-inch guns,
whose long muzzles, in pairs, project so far out of the turrets that hide their
mountings and firing-crew, let me tell you that it is 40 feet long, more than
4 feet in diameter, and weighs 60½ tons.   "It requires 550 pounds of
powder to load it, and the projectile weighs half a ton.   The muzzle-velocity
of the projectile is 2100 feet per second, with the stated charge, and its
energy is sufficient to send it through 26 inches of steel at a distance of
600 yards.   At an elevation of 40 degrees the range of the gun will be
not far from 15 miles."

In such a ship, deep down within the fortress is the massive and
complicated machinery, steam and electric, upon which the life and activity
of the whole structure depend.   The power is generated in four enormous

boilers, seventeen feet in diameter and twenty in length, their steel shells one and a half inches thick, built to carry a working-pressure of 160 pounds to the square inch.   Each pair of these boilers, placed fore and aft and side by side, is installed in a separate compartment, with fire-rooms at the ends.   Every boiler has four furnaces in each end, which give eight to each fire-room, or a total of thirty-two.   The two boiler compartments are separated by a water-tight bulkhead, and by a deep, broad coal-bunker.   At the sides of the ship are also coal-bunkers, which supplement the heavy armor-belt by the protection of a mass of coal twelve feet in thickness — in itself a not inconsiderable earthwork, which might arrest the fragments of a bursting shell that had succeeded in piercing the armor.   No casualty of naval combat can be worse than the penetration of high-pressure boilers by heavy shells.   Their complete protection is an imperative condition, quite as important as the protection of the magazines.

Such is a modern battle-ship — a "wonderful and complex instrument of warfare," as Lieutenant Staunton has expressed it.

> She is filled [he tells us] with powerful agencies, all obedient to the control of man — the creatures of his brain and the servants of his will. Steam in its simple application drives her main engines and many auxiliaries. Steam transformed into hydraulic power moves her steering-gear and turns her turrets. Steam converted into electrical energy produces her incandescent and search-lights, works small motors in remote places, and fires her guns when desired. Every application of energy, every device of mechanism, finds its office somewhere in that vast hull, and the source of all the varied forms of power lies in the great boilers, far down below danger of shot and shell, under which grimy stokers are always shoveling coal. Decades of thought and study, experiment and failure, trial again with partial success, and repeated trials with complete success, have assigned to each agency its appropriate function, and perfected the mechanism through which its work is performed.

These modern developments have added one entirely novel and tremendous adjunct to the fleet, in the torpedo-boat and its terrible weapon. These take the place to some extent of the fire-ship of a century ago, which was designed to injure the enemy not by silencing his guns or overcoming his gunners, but by insidiously destroying his ship itself.

The torpedo is, in its simplest form, simply some arrangement of a powerful explosive to be set off beneath or against the bottom of a ship, and shatter or sink it.   The idea is as old as gunpowder, but it is only in recent times that it has been made effective, — how effective we do not yet know.

Torpedoes are used in two ways: one is by fixing the torpedo beneath the water, either to be exploded by means of a percussion-cap when the ship runs against it, or from the shore by means of electricity.   Such arrangements as this, called submarine mines, are regarded as a most important means of defending harbors against hostile attack. During our Civil War they were extensively used by the Confederates, and were sometimes successful, as when one destroyed the monitor *Tecumseh* in Mobile harbor, during Farragut's famous attack there in 1864.

The former class, for which the word *torpedoes* is now reserved, includes explosive agents which are to be placed or sent against a ship's

THE MONITOR "TECUMSEH" SUNK BY A TORPEDO AT MOBILE, 1864.

bottom at sea and exploded there. Various devices of that kind, also, have been used for a long time in naval warfare. The Confederates tried hard to destroy several Northern vessels in the blockading squadron by devising very small, half-submerged boats, towing torpedoes astern, or else projecting on a long spar from their bows; and now and then they succeeded, as when one of the latter kind was made to sink the *Housatonic* off Charleston.

Then there have been invented, during the past fifty years, several cigar-shaped machines, which, by means of a chemical or compressed-air engine or clockwork, or some other application of power that might keep motive machinery within them going long enough, could be launched from shore or from another vessel and sent under water against a hostile ship. At first these were made to glide along just beneath the surface, carrying little flags that could be seen, and trailing two electric wires, enabling a

THE SEARCH-LIGHT REVEALING THE TORPEDO-BOAT.

person, by means of electric currents, to direct their flight; but latterly ingenuity has devised such an arrangement of rudders and self-acting balances within the torpedo's mechanism that it will continue perfectly straight upon the course it is aimed for, swerving neither right nor left, up nor down, and will explode the instant it touches an object hard enough to jar the delicate cap of fulminate in its snout. This latter kind, called the automobile (self-moving) torpedo, is now almost exclusively used, and some modification of the Whitehead is most popular.

It is cigar-shaped, and about twelve feet in length; the forward third is filled with gun-cotton— in quantity sufficiently powerful, if accurately applied, to ruin almost instantly the greatest battle-ship afloat.

All large war-ships are now fitted with tubes, opening near the water-line in various parts of the hull, which form gun-like exits for these terrible weapons, which are set in motion by a puff of gunpowder; but in addition to this every maritime government now has a number (Great Britain has more than 250) of small, swift steamers designed wholly for this purpose and called torpedo-boats. Most of them are a hundred feet or so in length, and intended to

A SELF-MOVING TORPEDO ON ITS WAY
TO ATTACK A MAN-OF-WAR.

accompany the fleet wherever it goes and in all weathers; but some are so small that they may be carried on the deck of a big cruiser.

All are made long, low, and narrow, and the speed of many of them exceeds thirty miles an hour. There is almost nothing to catch the wind or show above deck except a pair of short, flattened smoke-stacks, one behind the other; and the steersman stands, with only his head and shoulders visible, in a little box with windows that serves the purpose of a

A TORPEDO-BOAT AT FULL SPEED.

wheel-house. A mere wire railing saves the crew from sliding off the deck, and in action everybody stays below. No weight is carried that can be avoided, and the engines, taking steam from two boilers, are as powerful as can be packed into the space at command. Usually only coal enough for a few hours' steaming is carried, and every bushel of it is carefully selected as to quality, and is so treated and intelligently fed to the furnaces as to make the hottest possible fire, although never a spark must escape from the smoke-stack to betray the vessel in the darkness.

Next to speed the most important quality is ability to turn quickly, upon which might often depend the safety of the audacious little craft.

Torpedo-boats, however, are designed for a wider service than simply to

carry and discharge the frightful weapon from which they take their name.
They are to the navy what scouts and skirmishers are to a land army.
They form the cavalry of the sea, of which the cruisers are the infantry,
and the battle-ships and monitors the artillery arm. They must spy out
the position of the enemy's fleet, hover about his flanks or haunt his
anchorage to ascertain what he is about and what he means to do next.
They must act as the pickets of their own fleet, patrolling the neighborhood, or waiting and watching, concealed among islands or in inlets and
river-mouths, ready to hasten away to the admiral with warning of any
movement of the adversary.

It is not their business to fight (except rarely, in the one particular way),
but rather to pry and sneak and run, for the benefit of the fleet they serve.

ONE FORM OF SUBMARINE TORPEDO-BOAT.

But to insure all these fine
results, both officers and men
must be taught the art. Constant instruction and drilling
are necessary, and in each navy
a regular school of torpedo-
practice is maintained, where
the subject is studied in every
way. In the United States
such a school is kept at the
Newport (R. I.) Torpedo Station, where the torpedoes themselves are fitted for use and sup-
plied to the ships (the loaded war-heads are kept separately in the ship's magazine), and where one or more torpedo-boats are reserved for drilling purposes.

But a worse and more insidious foe than even these sneaking, hiding,
surface torpedo-boats threatens us in the submarine torpedo-boat, which
inventors have been experimenting with since naval warfare first began.
It is said that twenty-five hundred years ago divers were lowered into the
water in a simply constructed air-box, to perforate the wooden bottom of
an adversary's war-galley and sink it. Again, in our Revolutionary War, a
tiny walnut-shaped boat was made by an American, which was actually
tried. It would hold one man, and air enough for him to breathe for half an
hour. He would close the hatch, let in enough water to sink him a little
way, and then scull himself along by means of a screw-bladed stern-oar
until he got underneath the keel of an anchored vessel, to which, by ingenious
means, he would attach a can of gunpowder to be fired by clockwork, giving
him time to get away. It was actually tried and nearly succeeded. Robert

THE UNITED STATES BATTLE-SHIP "MAINE."

Blown up in the harbor of Havana, February 15, 1898.

Fulton, who made the first success of the steamboat, tried for years to contrive a submarine boat that would work, and succeeded so far as to scare British blockaders in 1812 very badly indeed; and the Confederates repeated the scare when the North was blockading their ports in the Civil War.

The great advantage of a submarine boat is, of course, its invisibility, and its safety from shot even if discovered; but the difficulties of progress and control as to depth and direction under water, and at the same time effective appliance of the explosive and safe retreat, are so many that they have as yet been only partly overcome. If the thing is ever accomplished, naval warfare will be demoralized until some adequate means be found to combat this unseen, destroying agency.

The principal agent in submarine attacks would probably be some form of dynamite, which, inhuman as its use seems, is slowly but surely taking its place among the weapons of war. The United States has one vessel primarily designed to employ dynamite, by hurling it in the form of shells. This volcanic craft is suitably named Vesuvius, and is a small, swift vessel having long tubes slanting upward through her forward deck, as shown in the illustration.

These tubes are the muzzles of great air-guns, through which she sends darts loaded with dynamite to fall upon a hostile ship or fort. It would not be safe, to say the least, to fire such bombs with gunpowder; and therefore pumps and engines in her interior compress air until it has acquired an expansive force sufficient for the purpose. When one of the darts has been laid in the breech of the tube, down beneath the deck, and suitably closed in, a valve is opened, the compressed air acts like burning powder, and away goes the dart, in a graceful curve to its target. In this case, of course, it is the vessel rather than the immovable gun that is aimed, and good marksmanship depends upon accurate calculation of distance; but remarkable shooting has been done. This system has never yet been tried in actual warfare, and may prove valuable chiefly in clearing harbors of mines.

CHAPTER VII

THE MERCHANTS OF THE SEA

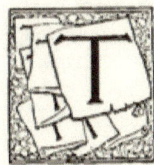HE history of shipping in an earlier chapter will also answer as a history of early international commerce. It began with the Egyptians and Phenicians, and was confined to their parts of the Mediterranean until after the middle ages, when it moved steadily to the western borders of Europe.

How great, rich, and influential were Tyre and its people we have already seen. A thousand years before the Christian era they controlled the commerce of the ancient world by reason of their wisdom as traders and their skill and energy as navigators and seamen. Turn to the twenty-seventh chapter of Ezekiel, and see how the Phenician metropolis was regarded, even in the time of that prophet, six hundred years before Christ. These Syrians had gradually extended their commerce until it took in the whole known world; and by their caravans to and from the interior of Arabia, Persia, India, and the Soudan, by their trains (perhaps of pack-horses) across Europe, by their marine expeditions to the Nile,—which they forced open to trade, for ancient Egypt was much like China in its exclusiveness,—and by their ships to all the Mediterranean ports, and up and down the Atlantic coast, they gathered and exchanged in the bazaars of Tyre and Sidon the products, manufactures, and luxuries of every country that had anything to sell. To the Phenicians, indeed, was ascribed, by the Latin and Greek writers of a few centuries later, the invention of navigation; and even when Phenicia had become of little account as a nation, its conquerors noted with admiration the skill of the men of that coast in seamanship. "They steered by the pole-star, which the Greeks therefore called the Phenician star; and all their vessels, from the common round *gaulos* to the great Tarshish ships,—the East-Indiamen, so to speak, of the ancient world,—had a speed which the Greeks never rivaled."

Later, in the days of the Roman supremacy, the trading-ships were as important to the country as its soldiers, for nearly every free man was in the army, and the slaves made poor farmers. A large part of the grain, as well as cattle, to supply the wants of the people, had to be brought from Egypt, which was pretty sure to have "corn," as the Bible calls it, when the rest of the world was suffering from short crops. Egypt supplied grain to Rome during the second Punic war, thus enabling her to resist the invasion of Carthage, and it is possible that Rome's later political alliance with Egypt was largely due to her interest in Egyptian crops. Large fleets of grain-ships, convoyed by armed vessels, were continually passing between the African coast and the Tiber, and so many were the risks they ran of wreck or capture, that the arrival of a flotilla with its precious freight of food was always a cause of rejoicing, at any rate, among the poor.

These merchant ships of classical times were broader and heavier than the war-galleys, and although they carried a few oars to help themselves in a difficulty, they ordinarily moved by means of sails, probably lugs. One of the grain-ships plying between Egypt and Italy about 150 A. D., according to Lucian, was one hundred and eighty feet long, slightly more than one fourth as broad, and forty-three and a half feet deep inside,— more like a barge than a "ship." The largest used in this trade would carry about two hundred and fifty tons. The transports that accompanied one of Justinian's fleets, A. D. 533, are stated to have carried one hundred and sixty to two hundred tons of supplies each.

These Roman vessels were made of pine, and were coated with a composition of tar and wax, then painted, often with elaborate decorations in bright colors, with pigments mixed with melted wax. Now and then one was built of truly vast proportions, as that one which brought from Egypt to Rome the first of the stolen obelisks.

With that grand awakening of interest in education, industry, and discovery which took place in the fourteenth century, the city of Venice gained the lead in power, and her merchants became the most enterprising and wealthy. It was the expansion of commerce that urged the explorations that marked the fifteenth and sixteenth centuries, for by this time Venice had her banks—the first in the world to approach the character of modern banks—and her exchange on the famous Rialto bridge; Genoa was in close rivalry; Spain was gathering immense quantities of gold in South America; and England was coming to the front as a maritime power. The trade with Cathay—as India, China, and the Oriental islands were called collectively—was chiefly by caravans across the Persian deserts, and Spain, England, and Holland had small shares in it, since the only water-route known was through the Mediterranean and Red seas, where, between the perils of the ocean, the extortionate charges and stealings of the Arabs (who carried the cargoes from vessel to vessel across the Isthmus of Suez), and the risk of capture by Algerian pirates, there was little chance left for profit to either merchants or ship-owners.

To western Europe, then, Vasco de Gama's discovery of the route around the Cape of Good Hope was a long advantage, and England and Holland at least were quick to seize it. The great "East India Companies" of the Dutch and English were formed by a group of powerful merchants in London and in Amsterdam, who were given vast privileges by their governments in respect to trading in the East. The Dutch company was not founded until 1602, two years after the English company, but it soon became the more prominent of the two, and was one of the principal means by which the Netherlands secured the preponderance of the carrying trade of the world, bringing to her ports, by the middle of the seventeenth century, almost all the commerce previously enjoyed by Cadiz, Lisbon, and Antwerp, and making very serious inroads upon that of London and Bristol. The Dutch East India ships, copied from the

A CAPTAIN IN THE MERCHANT MARINE.

Genoese carracks, were the biggest merchant vessels then afloat, well able to cope with many of the war-ships; and two hundred of them were at this time engaged in the Asiatic trade alone.

It was in aid of the English rival company not only, but as an attempt to save and revive the commercial position of England generally, that Cromwell's "navigation laws" were enacted, prohibiting the carriage of goods to or from British shores except in ships owned and manned by Englishmen,—laws that were aimed directly at the Dutch, and led to the

A CLIPPER ESCAPING FROM THE "ALABAMA."

long wars of the latter half of the seventeenth century. These were called wars for the supremacy of the sea, but actually they were a prolonged struggle for the biggest share of the world's trade, which is the only real value of the "supremacy of the sea." It is a saying that "trade follows the flag," and so it does; but at the beginning the flag goes were the trade is to be had.

These companies were so mixed up in the politics of their respective governments that it would be a long task, although entertaining, to trace their growth, which is really that of western civilization in the East. They equipped fleets of merchant and war vessels, established forts, carried on small wars along the Oriental coasts, and were really little kingdoms within kingdoms, because of their wide monopoly, enormous wealth, and the national importance of all their enterprises. The final result was that, as Great Britain finally overcame the Dutch and French at home, so her East India Company ousted them from India; but it was not until 1858 that old "John Company," which had come to be regarded by the natives of India as the government itself, was dissolved, and resigned its territories to the crown and a system of trade open to all the world.

Those were slow and costly times compared with the present, though seeming to us full of a romance impossible now. A voyage around the world occupied three years, and to go from London to Calcutta and back took from New Year's to Christmas under the most favorable circumstances. Another important change, too, has gradually come about. Formerly, the vessels were owned almost entirely by the merchants themselves, or by a company of them; they paid all a ship's expenses, and put into her a cargo of their own wares. They would send to China, for instance, cotton goods, household furniture, hatchets, tools, cutlery and other hardware, farming implements, and fancy goods of all sorts. In return the vessels would bring silks, tea, and porcelain, which would go into the owners' warehouses and be sold in their own shops. Shipper, importer, and merchant were all one.

Now this is changed. The importers and merchants of London, Hamburg, and New York are not often those who own vessels and bring their own goods. Instead of this they have agents, who live permanently in each of the foreign ports, where they buy the merchandise they want and hire a vessel, or the needed space in a vessel, belonging to somebody else to bring them home. By the old way, the nation which had anything to sell carried it to the nation that would buy it, and brought back the best thing it could get in exchange; now the merchants go to various parts of the world, buy their cargoes, and order them sent home, in substantially the same way as you go a-shopping in town.

This has brought out a new department of sea-labor, unknown, as a class, a century ago — the business of carrying goods which the owners of the vessels have no property in. In London, New York, Hamburg, and all other seaboard cities of this and other countries, the great majority of the shipping is owned, not by the merchants of the city, but by "transportation companies," who agree to carry cargoes at a certain rate.

Merchant vessels may be divided into three classes, of which the first includes steamships and sailing-vessels planned primarily for freight trans-

THE SALOON OF A SAILING PACKET-SHIP, ABOUT 1840.

portation, which run back and forth between certain ports, and so constitute "lines" for freight. Such lines exist along even the remotest coasts, so that goods may be shipped directly, or by a single transfer, from any given seaport to almost any other in the world. Some of these lines, sailing between certain ports, are devoted to particular uses, such as those of oil-steamers and cattle-steamers. The oil-steamers run between America and Europe with American petroleum, and in the Black Sea and the Mediterranean with oil from Russia; the entire holds are divided into vast iron tanks for this liquid, which is poured into and pumped out of them as into and out of a great barrel. The cattle-steamers are specially arranged for the transportation of live stock, but one line, running between America and England, also carries passengers at a cheap rate. The second class of vessels consists of those which make the transportation of passengers their first object, loading their holds with first-class freight, for which high rates are paid in consideration of its swift delivery. The third class includes what are known as "tramp" steamers, which run irregularly, as the old sailing-vessels used to do, picking up cargoes wherever they find them and carrying them to any port. They are often of great size and power, but being under less close supervision are often less careful as to the safety of crews and cargoes, and are sometimes unseaworthy. They are always ready to answer any sudden demand for ships, their owners keeping watch of the chances and telegraphing to their captains where to go for their next cargoes. Without the submarine telegraph these tramp steamers could scarcely compete with the regular lines; but, besides the great transoceanic cables, all the sea-coasts are now festooned with electric cables, which have frequent stations and connect the important ports of America and Europe with those of Africa, Persia, India, the Spice Islands, Australia, and New Zealand, and there is now a plan to run a cable across the Pacific between America and New Zealand, by way of the Sandwich Islands, Samoa, and Fiji.

A CORNER IN THE SALOON OF A MODERN STEAMSHIP.

The passenger-ship is a distinctly modern feature of marine carriage. In former days the few persons who were obliged to cross the seas on business errands, and the fewer who went abroad for health or pleasure or the love of travel, had to accept such rough accommodations as the ordinary merchant ships afforded. But as soon as the East and West Indies were added to the map of the world, and colonies of Europeans began to settle on distant coasts and islands, the amount of travel justified owners of vessels in enlarging cabins and providing comforts likely to induce patronage of their lines. Even two hundred and twenty-five years ago the voyage between India and England around the Cape of Good Hope, though it became somewhat tedious, because it lasted six or seven months, was by no means a miserable experience in a well-found ship. Thus Dr. John Fryer has recorded of such a sea-journey in 1682 that "it passed away merrily with good wine and no bad musick; but the life of all good company, and an honest commander, who fed us with fresh provisions of turkies, geese, ducks, hens, sucking-pigs, sheep, goats, etc."

A century later, when England had come firmly into possession of India, and thousands of her officers, troops, and traders, with their families, were colonizing her ports, there were demanded the largest and finest ships that could be built, combining accommodations for many passengers with great cargo capacity. Such were the great East Indiamen; and in those leisurely days a trip half-way round the world on one of these roomy old vessels was a continuous pleasure to almost every one that undertook it.

> The ship was a bit of Old England afloat, where the passenger rented for so many months a well-lighted, roomy, unfurnished apartment, which, according to his taste and means, he fitted up for the voyage with numberless comforts and sea stores that none but a yachtsman would think of cumbering himself with at sea to-day; and, reading narratives of the old long sea-voyages, one is constantly coming across expressions of regret by passengers when they "took leave of the good ship that for so many months had been their floating home." These fine old passenger sailing-ships were, like a man-of-war, entirely dismantled at the end of each homeward voyage, and underwent a complete overhaul and refit before starting out again on an outward one. Passengers usually sold their state-room furniture by auction on board the ship on her arrival in port.

Such a ship, the Atlantic packets, and even men-of-war bound on a long blockading cruise, did not hesitate to stow aboard all the live stock that room could be found for, sometimes by comical devices. In that book of charming reminiscences of ways and means afloat before the days of quick steam transit, "Old Sea Wings," Mr. Leslie has a chapter which he calls "The Old Ship-Farm," where one may learn curious particulars of this matter.

> The man in charge of this part of the stores was the ship's butcher, and he had as "mate," or assistant, a youth of all work known to all sailors as "Jemmy Ducks." Their barn, or storehouse, was especially the great long-boat, which often looked more like a model of Noah's ark than a craft serviceable in case of shipwreck.
> Always securely stowed amidships, well lashed down and housed over, the boat, as she lay upon the ship's deck, was full of live provender, being divided, as to her lower hold, into pens for sheep and pigs, while upon the first floor, or main deck, quacked ducks and geese, and above

them (literally in the cock-loft) were coops for another kind of poultry. This great central depôt was closely surrounded by other small farm-buildings, the most important being the cow-house, where, after a short run ashore on the marshes at the end of each voyage, a well-seasoned animal of the snug Alderney breed chewed the cud in sweet content. In fact, when, in the old days, a passenger-ship began her voyage, the hull of her clumsy long-boat was nearly hidden by the number of temporary pens and sheds required to house the live stock for the supply of her cabin table; and with its many farm-yard and homelike sounds a ship was, even then, more like a small bit of the world afloat than it is now.

There was always regular traffic between America and Europe, especially with Great Britain, and the rapid growth of emigration to the United States and Canada made it profitable, early in this century, to put on fast-sailing packet-ships, making voyages, at intervals of a month, between London and New York. By 1840 a man might find a large, well-ordered ship departing every week or so for the transatlantic passage, which usually required less than a month going east, but might be two weeks longer coming west. Their cabins were as comfortable and perhaps more homelike than any seen now, and quite as pretty, with their white and gold paint, cut-glass door and locker knobs, damask hangings, dimity bed-curtains, and other old-fashioned niceties; and the fare was abundant and varied, as it ought to be in a neat ship with a small dairy aboard, and perhaps a green-salad garden planted in the jolly-boat. None of these packets were more popular than those of the well-remembered Black Ball Line.

The steerage passengers were not so well off then, though they seemed to stand the voyage quite as well as nowadays. The fare was twenty-five dollars, and the passenger found himself "in everything but fire and water." "Steerage passengers then had to cook their own victuals, weather permitting, at an open galley-fire on the waist-deck; . . . but in anything like rough weather, all steerage passengers had either to run the chance of getting constantly wet with salt water or keep below." The 'tween-decks space allotted to them was almost completely filled by rows of bunks, built in each port by the ship's carpenter, in three tiers, one above the other, though the ceiling was scarcely seven feet from the floor; and when in a stormy time the hatches were closed the only way the crowd could find room was by most of it stowing itself away in the bunks, while a few tried to sit or lie on the luggage piled in the narrow aisles. The only light was that of a few candle or whale-oil lanterns, and in a very bad storm everybody came near smothering, for then it was impossible to ventilate the steerage properly without flooding it. Considering that all the provisions for the steerage people were kept in this crowded, damp, and fearfully close room, it is marvel-

FAIR WEATHER ON THE DECK OF A CLIPPER-SHIP CARRYING GOLD-SEEKERS TO CALIFORNIA IN 1849.

ous that a pestilence did not break out during every voyage, but, in fact, sickness was rare.

The introduction of steam into oceanic navigation was experimented with as soon as river steamboats were successfully built. The first vessel to go across the ocean by the aid of a steam-engine is said to have been the *Savannah*. This vessel, built in Savannah, Ga., and having a steam-engine and paddle-wheels, certainly crossed to Liverpool in 1819; but it is asserted that she sailed all the way, using her steam very little, if at all, although making the trip in twenty-two days. In 1825 the English steamer *Enterprise* went from London to Calcutta; but it was not until some years later that ocean navigation by steam became successful in the beginning of operations by the Cunard Company in 1833.

These first steamers were side-wheelers, and their huge boilers and simple engines consumed so much fuel that the space taken up by the coal, added to that devoted to passengers, left little room for cargo. Moreover, their speed was less, often, than that of the "clippers," so that for some time the sailing-packets maintained their competition. The adoption of the screw propeller, in place of the costly and cumbersome side-paddles, and the perfection of the compound marine engine, which effected a great saving in fuel, soon established the superiority of steam navigation for passenger service, fast freights, and service in war,—yet even these improvements were not fairly brought about until the first half of the present century had gone; and sails are not yet abandoned, not only because they steady a vessel in a gale, and may help her decidedly when the wind is fair, but may save her altogether in case of the disabling of her machinery.

Great modifications and improvements on old models have grown out of the employment of steam and the screw, and human invention has been taxed to the utmost to combine economy of space and expense with the various needs of different climes, or special cargoes, or the demands of a traveling public that is growing more fastidious every day. The most obvious changes in naval construction have been in the greatly elongated hull, the enormous dimensions aimed at, and the all but universal employment of iron. When the first steamship crossed the ocean the proportions of ships averaged three to five beams in length. . . . But it was discovered that with a given power and depth and beam the length could be increased without materially affecting the speed, thus adding to the carrying capacity of steam. Great length to beam, however, does not necessarily imply great speed; the speed of beamy vessels has too often been demonstrated. Fineness of lines is equally essential, together with the proper distribution of weights, and the like. The great average speed exhibited by the modern steamship is due in large part to the momentum of such a vast weight, which, once started, has a tremendous force.

Long after the transatlantic steamships were regularly running, sixteen or seventeen days was considered a good passage between New York and Liverpool. Then the Inman and White Star lines began to see the importance of faster speed, and their rivalry had cut this estimate in two by 1870, and ten years later the Guion Line's *Arizona* and other crack boats took a full day off that. Since then there has been a steady improvement in speed, as is shown by the table below; and this seems to have followed proportionately the steady increase in length. The ships of 1850 never reached 300 feet in length, and few were over 2300 tons in burden measurement. By 1880 almost all the first-class "liners" of the world exceeded 450 feet, and some soon approached 600, as the *City of Rome* (586 feet, 8826 tons), and several of the famous Hamburg liners, White Stars, and Cunarders nearly equaled her in dimensions (*Paris* and *New York*, 580 feet each; *Teutonic* and *Majestic*, 582 feet); while some of the more recent boats are even longer, as *Campania* and *Lucania*, 620 feet, and the gigantic *Kaiser Wilhelm der Grosse*, 648 feet. Two other ships, now planned, will considerably exceed this length. The total number of transatlantic passenger-steamships regularly sailing from New York alone is now between 90 and 100, belonging to 14 different lines. The table of speed-records between New York and Queenstown, since the time was reduced to less than six days, is as follows:

Year.	Steamer.	Line.	Direction.	Date.	Days.	Time. Hours.	Min.
1882	*Alaska*	Guion	Eastward	May 30 to June 6	6	2	0
1891	*Majestic*	White Star			5	18	8
1891	*Teutonic*	White Star	Westward	Aug. 13–19	5	16	31
1892	*Paris*	American	Westward	Aug. 14–19	5	14	24
1893	*Campania*	Cunard			5	12	7
1894	*Lucania*	Cunard	Westward	Sept. 8–14	5	8	38
1894	*Lucania*	Cunard	Eastward	Oct. 21–26	5	7	23

The approximate distance between Sandy Hook (light-ship), New York, and Queenstown (Roche's Point) is 2800 miles. The fastest day's run on record, however, was made by the *Kaiser Wilhelm der Grosse*, of the Nord Deutscher Lloyds Line, averaging 22.35 knots (or nautical miles, of 6080 feet each) per hour, equal to about 25½ land miles. From Sandy Hook to Queenstown deduct 4 hours 22 minutes for difference in time. Queenstown to Sandy Hook add 4 hours 22 minutes for difference in time.

This eager rivalry in respect to speed, which insures not only a larger and more influential passenger service, but increased business in fast freight and in the carriage of mail — both highly remunerative — is only one feature of the sharp competition between these ocean carriers as to which shall offer the greatest advantages, and this is of benefit to the public, though it has not greatly cheapened fares.

Men travel far more now than they were wont in the time of "good Queen Bess," or even of our own grandfathers, and the few travelers for

EMIGRANT PASSENGERS EMBARKING UPON A TRANSATLANTIC "LINER."

pleasure of those days would scarcely believe their eyes if they could look
into the floating palaces — almost cities — in which we brave old ocean now.
A ship of one of the better passenger lines is a little world in itself, containing almost all the appliances of the best modern hotels on shore, and reducing the inevitable inconveniences of life on shipboard by clever devices of
every sort. In the one matter of ventilation the ingenuity of the builders is
particularly taxed. Money is spent lavishly in the finishing and furnishing
of these great ships, not to mention the expense of running them, which
sometimes amounts in cost of fuel, food, and wages to $5000 a day.

The steamship lines between New York and Great Britain do not steer
straight across the Atlantic, but on their way to this country keep well to
the northward, so as to get to the west of the Gulf Stream, and into the
favorable current flowing south from Baffin's Bay; then they skirt Newfoundland, Nova Scotia, and Cape Cod. Going east, however, the steamers — and sailing-vessels too — keep farther south, and work along with the
Gulf Stream as far as they can. From Europe to South America, or
through the Straits of Magellan on their way to the South Sea islands or
Australia (though this route is not often taken), or to the Pacific coast of
the Americas, vessels keep close down the African coast, and then steer
straight ahead from Guinea to Brazil, and on down the American coast.
(Put a map before you and you will understand these courses better.)
Sailing-vessels to Europe or the United States from Cape Horn, however,
would swing far out into the South Atlantic to avoid heading against the
southward coast-current and to get the benefit of the southwest trade-wind
and the equatorial currents. Between New York and the Cape of Good
Hope the track is nearly straight.

In the Pacific, the steamer-route between San Francisco or Vancouver
and China and Japan, instead of being as direct as a parallel of latitude,
takes a southerly course when bound west, and a northerly course when
bound east, the exact lines varying with the seasons as the prevailing winds
and currents change. What these winds and currents are is explained in
another chapter; but it is interesting to note that there is a difference of
many miles in the ordinary westerly and easterly courses, the latter being
much the shorter, although the vessels of the Canadian Pacific Line often
sail so far north with the Japan warm current as to sight the Aleutian
Islands. Sailing-vessels, moreover, curve so much farther south than
steamers in going west from San Francisco, in order to take advantage of
the equatorial current and the trade-winds, that the space is a thousand
miles north and south between ships outward bound and those coming
home. Between California and Honolulu a steamer takes a bee-line, but

A "WHALEBACK" FREIGHT STEAMER, ALSO ADAPTED TO PASSENGER SERVICE.

sailing-vessels find it best to make detours. In summer, when outward bound, this amounts to steering straight northward until under latitude forty degrees, before turning westward, making an angular course that looks very unnecessary to a landsman.

I have said that the finding of a sea-route to the East around the Cape of Good Hope was a great boon to western Europe, and advanced commerce. It remained so until within the last seventy-five years. Lately, the corsairs being out of the way, and safety guaranteed in Egypt, merchants and sailors both began to wish they had a shorter route between England and India. Then, with immense labor and sacrifice, the canal was cut across the Isthmus of Suez, and commerce returned to its ancient channel through the Red Sea, saving thousands of miles of weary distance.

From the end of the Red Sea at Aden, the tracks of steamers both ways are straight courses to Bombay or Ceylon, and thence right up to Calcutta, across to Singapore, or down to Australia. Except East African coast lines, few steamers go around the Cape of Good Hope from England, excepting one line to South Australia, which steers straight eastward all the way from Cape Town to Adelaide, 6125 miles. But the Indian Ocean is so situated under the equator, is so filled with prevailing winds and currents and counter currents, that sailing-vessels must take very roundabout courses there, and can by no means steer the same track at all seasons of the year. These voyages from New York and London to the East are the longest regular sea-roads. A short table of distances between well-known ports along regular steamer-routes will be of interest; and by reversing them, or adding them together, the sailing distance between almost any two ports on the globe may be calculated.

	MILES.		MILES.
Acapulco to San Francisco	1,850	Liverpool to New York	3,057
Aden to Bombay	1,635	Liverpool to Para	4,010
Aden to Colombo (Ceylon)	2,100	Liverpool to Quebec	2,634
Aden to Zanzibar	1,770	Marseilles to Algiers	410
Auckland to Honolulu	3,915	Montevideo to Magellan Strait	1,070
Auckland to Suva (Fiji)	1,140	New Orleans to Havana	570
Cadiz to Teneriffe (Canaries)	698	New York to Colon	1,980
Cape Horn to Rio de Janeiro	2,350	New York to San Francisco, about	17,000
Cape Town to Plymouth (Eng.)	6,016	New York, via St. Thomas, to Para	3,130
Cork to St. John's (N. F.)	1,730	Panama to San Francisco	3,260
Ceylon to West Australia	3,305	Porto Rico (San Juan) to Havana	1,030
Glasgow to New York	2,790	Rio de Janeiro to Plymouth	4,911
Havre to Martinique	3,560	San Francisco to Honolulu	2,080
Havre to New York	3,160	San Francisco to Yokohama	5,280
Hobart (Tas.) to Invercargill (N. Z.)	930	Shanghai to Yokohama	1,033
Hong Kong to Manila	650	Singapore to Hong Kong	1,430
Hong Kong to Shanghai	800	Suez to Aden (length of Red Sea)	1,308
Hong Kong to Yokohama	1,620	Suva to Honolulu	2,783
Leith (Scot.) to Iceland	1,050	Sydney to Auckland	1,281
Lisbon via Dakar (W. Af.) to Pernambuco	3,297	Sydney to Vancouver (B. C.)	6,780
Lisbon to Cape Verd Islands	1,537	Teneriffe to Porto Rico	2,790
Liverpool to Barbadoes	3,646	Trieste to Bombay	4,317
Lisbon to Para	4,000	Yokohama to Honolulu	3,445
Liverpool to Lisbon	983	Yokohama to San Francisco	4,750
Liverpool to Madeira	1,430	Yokohama to Victoria	4,320
Liverpool to New Orleans	4,767	Zanzibar to Bombay	2,400

CHAPTER VIII

ROBBERS OF THE SEAS

S the sea has furnished opportunities for so much good,— for manly exertion, knowledge of the world, and acquaintance with people outside of one's own country, and for gaining wealth,— so it has given a chance for unscrupulous men to show the worst that is in them; and the guarding of shore towns and merchant vessels from piratical attacks has always been a part of the usefulness and duty of a nation's naval force.

As on land there are robbers and highwaymen, so on the ocean robber ships have often been lying in wait for vessels loaded with treasure, and have landed crews of marauders to make havoc with rich seaboard provinces. Such robbers on the high seas are termed pirates, and their crime was visited by the old laws with torturing punishments; yet they were never more daring than when the laws against them were severest.

The word is Greek, and the first pirates who figure in history are those of the Greek and Byzantine islands and coasts — bloody ruffians who originated the amusing method of disposing of unransomed prisoners by making them "walk the plank," as has been done within the present century.

The intricate channels and hidden harbors of the Ægean Sea long remained a hiding-place of sea-robbers, and are still haunted by them, though every few years, from Cæsar's time till now, the kings of the surrounding countries have sent expeditions to break them up. In the sixteenth century piracy in that region was especially prevalent. The crews then were chiefly Turkish, but the great leaders were two renegade Greeks, the brothers Aruck and Hayradin Barbarossa ("Redbeard").

It happened that Spain, having conquered the Moors of Granada in 1492 and pursued her victories across the straits, had gained control of Algeria (at that time a collection of small Mohammedan states), and held it until the death of King Ferdinand in 1516. Then the Algerians sent an embassy to Aruck (sometimes spelled Horuk, or Ouradjh) Barbarossa, requesting him to aid them in driving out the Spaniards, and promising him a share in the spoils. He eagerly accepted this proposition, seeing a great deal more in it than the Algerians saw; and the moment the Spaniards had

WALKING THE PLANK.

been beaten and expelled he murdered the prince he had come there to help, seized upon the city and port for himself, and made it the headquarters of that system of desperate piracy which became the dread of all Europe. These robbers of the sea called themselves *corsairs*, from an Italian word signifying "a race"; and they generally won, because they had the best and swiftest vessels of that time, such as feluccas, xebecs, and the like. The black flag which they flew was not blacker than their reputations, so that even yet to call a man as bad as a Barbary pirate is to mean that he could not be much worse if he tried. The Spanish colonies in America, a few years later, began sending home immense treasures dug in the silver- and gold-mines of Peru and Mexico, and extorted from the natives or stolen from the temples of those unhappy countries. A quantity of ingots and gold and silver ornaments equal in value to fifteen million dollars of our modern money was taken at one time by Pizarro, in Peru, as the ransom of the Inca Atahualpa, and booty amounting to a similar sum was gained in the sacking of various cities. This great inpouring of wealth caused a general giving up of manufactures and trade in Spain, and was one of the reasons of her final decline in power, and it had the immediate bad effect of making piracy more attractive than ever. The treasure-ships, though convoyed by war-ships, were often attacked and captured by the corsairs. Barbarossa's fleets were more like armadas of a powerful nation than mere pirate craft; and whenever it happened that his commanders were defeated, they would land upon the nearest unprotected coast of Spain, France, or Italy, and pillage and burn some town in revenge. How galling this was to all merchants and travelers we can hardly understand in these days; but so strong were the corsairs that the fleets and armies of various governments, and even of the Pope, which were sent against them, could not gain their stronghold nor suppress their cruisers, at least for more than a short time. Charles V of Spain tried greatly to conquer them; but although his forces, attacking Aruck Barbarossa from the province of Oran, near Algiers, defeated and killed him, Hayradin (more properly spelled Khair-ed-din) Barbarossa succeeded his brother, and, placing himself under the protection of Turkey, continued to build up the power of the pirates. His first care was to fortify the city of Algiers, and he expended a great deal of money and labor on the perfection of the harbor, compelling all his prisoners and thousands of citizens to work as slaves on the defenses. Next he conquered Tunis, and was selected by the sultan as the only fit man to sail against Andrea Doria, the great Genoese naval commander of the Christians in their wars against the Turks early in the sixteenth century. Mediterranean commerce became so unsafe that watch-towers were built all along the coasts, and guards were kept afoot to give alarm at the approach of the corsairs. Charles V gathered together a powerful armament, and sailed to the rescue of Tunis, recapturing it for its rightful sovereign in 1535; but he was never able to capture Hayradin Barbarossa, who lived out his life in Algiers as "a friend to the sea and an enemy to all who sailed upon it."

After his time the power of the pirates continued under other leaders; and not Algeria alone, but Tripoli, Morocco, and even Tunis, harbored piratical vessels in every port, and the rulers shared their spoils; piracy, indeed, was the source of their national revenues, and was encouraged by the Sultan of Turkey inasmuch as all these states were his vassals.

Every few years some European power—Spain, France, Venice, or England—would lose patience, send a fleet, and open a campaign that would be successful in destroying certain strongholds, releasing a crowd of prisoners, and burning or sinking many ships. The city of Algiers was bombarded almost into ruins in 1682, and the job completed a year later, after the Algerians had tossed the French consul out to the fleet, with their compliments, from the mouth of a mortar. They were fond of such jokes. Nevertheless, the city speedily recovered, and piracy, complicated by Moslem fanaticism and Turkish politics, harassed commerce during all the next century, partly because Europe was so busy in its own wars that it had no time for outside matters, and partly because it was for the advantage of certain nations (particularly of Great Britain, which, in possession of Gibraltar and Port Mahon, might have suppressed this villainy) to let the corsairs prey upon its foes—especially France. The actual result was that most or all of the European powers fell into the custom of paying to Algiers, Tunis, Tripoli, and other rulers of the Barbary (or Berber) States large sums of money as annual tribute to restrain them from official depredations upon their coasts and commerce, besides other large payments for the ransom of such Christian prisoners as each sultan's lively subjects continued to take in spite of treaties.

In this shameful condition of affairs the newly independent United States was obliged to join during the first years of its existence, to secure immunity for our commerce in the Mediterranean, because we had not yet had time to create a navy. By the end of the century, however, the United States was able to defend itself at sea, and in 1801 answered the insults of Tripoli by bombarding its capital seaport until the dey sued for mercy and promised to behave himself. Nevertheless, he needed another lesson, and in 1803 a second American fleet was sent to the Mediterranean, commanded by Preble, in the *Constitution*, with such subordinate officers as Bainbridge, Decatur, Somers, Hull, Stewart, Lawrence, and others that later became famous. One incident of this campaign, which began by frightening the Sultan of Morocco at Tangier into abject submission, but was especially directed against Tripoli, is well worth remembering.

Captain Bainbridge, going alone in the fine frigate *Philadelphia* into the harbor of the city of Tripoli, had unfortunately run aground, and there, overpowered by the number of his enemies afloat and ashore, had been compelled to give up his ship, and find himself and all his crew taken prisoners. He managed to get word of his misfortune to Commodore Preble at Malta, and that officer at once took his fleet to Tripoli — Decatur, in the

THE "ARGUS" CAPTURING A TRIPOLITAN PIRATE FELUCCA.

Argus, gallantly capturing on the way one of the great lateen-sailed piratical crafts of the enemy, which later proved a useful instrument in the contest. The fleet blockaded Tripoli for a while, and shelled the fortifications somewhat, just to give the bashaw a hint, and to encourage the poor prisoners; but none of the big vessels was able to enter the narrow, tortuous, and ill-charted harbor in the face of the many batteries, under whose guns the *Philadelphia* could be seen at anchor with the Tripolitan flag at her

main, so they sailed away to Syracuse to make preparations for reducing
this nest of barbarians. Gunboats of light draft and mortar-vessels had
to be fitted out; but the first thing was to try to carry out a plan that
Decatur and all his friends had been maturing ever since they had arrived—
the destruction of the *Philadelphia*, not only because she had been refitted
into a powerful weapon in the hands of the enemy, but because it was galling to national as well as naval pride to see her flying a foreign flag. The
plan was this:

Decatur was to take a picked crew of seventy officers and men on the
captured felucca (renamed *Intrepid*), and attempt at night to penetrate to
the inner harbor of Tripoli in the disguise of a trader, supported as well as
possible by the gun-brig *Siren*, also disguised as a merchantman. As his
pilot was an Italian and a competent linguist, it was hoped the ketch
could get near enough to set fire to the ship, whirl a shotted deck-gun into
position to send a shell down the main hatch and through her bottom, fire
it, and escape before the surprise was over. The chances of failure were
enough to daunt the bravest, yet every man in the fleet wanted to go.

On February 15, 1804, Decatur in his felucca, and Somers commanding
the brig, found themselves, toward evening, again in sight of the town, with
its circle of forts crowned by the frowning castle. The great *Philadelphia*
stood out in bold relief, closely surrounded by two frigates and more than
twenty gunboats and galleys. From the castle and batteries 115 guns
could be trained upon an attacking force, besides the fire of the vessels, yet
the bold tars on the *Intrepid* did not quail.

The crew having been sent below, the pilot Catalona took the wheel,
while Decatur stood beside him, disguised as a common sailor. It was now
nine o'clock, and bright moonlight. Standing steadily in, they rounded to
close by the *Philadelphia*, and, boldly hailing her deck-watch, asked the privilege of mooring to her chains for the night, explaining that they had lost
their anchors in the late storm, and so forth, until at last consent was given.

Having dragged themselves close to the frigate, it was the work of only
a moment to board her with a rush, overpower her surprised crew, and make
sure of her destruction by means of the combustibles and powder they had
brought with them. Before their task was done, however, they had been
discovered, and it is almost a miracle that they were able to return to their
felucca, and make their way out of the harbor, through a rain of harmless
cannon-balls; yet they did so, and Decatur was justly honored for one of
the most gallant exploits in naval annals.

A few weeks later Preble's squadron shelled the pirate city and fortresses
into ruin, forced Tripoli as well as Algiers and Tunis to respect then and

thenceforth the American flag, and gave these arrogant rulers the new sensation of paying instead of receiving money for bad deeds. It put an end to the corsairs.

Turkish and Barbary pirates were not the only ones in the world, however. Although the old Norwegian vikings and rough Norman barons did not go under that name, they were scarcely anything else, in fact, as the neighboring peoples could testify, though this was far back before modern history began. But when the Spaniards and the French began to colonize the West Indies, and to dig mines in South and Central America, not only were the Barbary corsairs given a fresh incentive, but a new set of pirates sprang up, the most daring that the world has ever seen.

As the archipelago east of Greece had sheltered the hordes of the Turkish sea-robbers, so the many islands, crooked channels, reefs unknown to all but the local pilots, small harbors, and abundant food of the Antilles, made the West Indies the safest place in the world for pirates to pursue their work. To these new and wild regions, in the sixteenth century, had flocked desperados and adventurers from all over the world. When the wars with their chances of plunder died out after the campaigns led by Cortés, Pizarro, Balboa, and the rest of the Spanish *conquistadores*, many ruffians seized upon vessels by force, or stole them, and turned into robbers of the sea. At first, as a rule, they had farms and families on some island, and went freebooting only a portion of the year. The island of Hayti, or Santo Domingo, was then settled by farmers, hunters, and cattlemen, the last-named of whom, mainly French, passed most of their time in the interior of the island, capturing, herding, or killing half-wild cattle and hogs. But the monopolies which Spain imposed upon the colonists interfered with the market for their produce and induced an illicit trade, which led to frequent encounters with the Spanish navy. As the constant wars between Spain and France and England increased the difficulties of trade, large numbers of the colonists joined the freebooters, who then became extremely numerous and formidable, losing their old name and becoming known by that of the cattlemen — buccaneers, from the French word *boucanier*.

First Santo Domingo, then Tortugas, and finally Jamaica were headquarters of the buccaneers, who were made up of men of all nations, united by a desire to prey upon Spain as a common enemy. They were thousands in number, possessed large fleets of ships and boats, were well armed, and finally formed a regular organization with a chief and under-officers. The most noted of these chiefs, perhaps, was Henry Morgan, a Welshman, who was at one time captured and taken home to England for trial. To his own surprise, instead of being executed, he was knighted by Charles II, who had not been at all grieved at seeing Spanish commerce harassed; and Morgan was returned to Jamaica as commissioner of admiralty, where at one time he acted as deputy governor, using his opportunity to make it unpleasant for those of the buccaneers with whom he had formerly had disagreements as to the distribution of prizes.

The earlier buccaneers found ample plunder in the Spanish fleets. They patrolled the sea in the track of vessels bound to and from Europe, and seized them, allowing or compelling the crews to become pirates, or else to be killed or carried into slavery. This work, however, employed only a portion of the buccaneers; and early in the seventeenth century, as the commerce of Spain declined, it became too uncertain a means of wealth to suit them. But the rich Spanish settlements still remained; and often, therefore, they equipped a great fleet, enlisted men under certain strict rules as to sharing the spoils, and sailed away to pillage some coast. There was hardly an island in the West Indies from which, in this way, they did not extort immense sums of money under threat of destruction of the people. The mainland also suffered from the marauders. Great cities, like Cartagena in Venezuela, Panama on the Isthmus, Mérida in Yucatan, and Havana in Cuba, were attacked by armies of buccaneers numbering thousands of men. Sometimes their fortifications held good, and the enemy was beaten back; but sooner or later all these cities, and others, smaller, were captured, robbed of everything valuable that they contained, and burned or partly burned.

For years the buccaneers were the terror of the Caribbean region, and after the famous sacking of Panama, under Morgan, in 1671, their power spread across the Isthmus and scourged the southern seas. We have no way of knowing the amount of the treasure which they captured from the merchant vessels and from the coast of Peru; for the moment they got home from an expedition they wasted all their booty in wild carousing, so that the spoils earned by months of exposure, and wounds, and danger of death, would be spent in a single week.

At last even England and France, after secretly favoring the buccaneers, became roused to the necessity of controlling them, and it was with this object in view that a certain Captain William Kidd was fitted out at private expense toward the end of the seventeenth century, and armed with King William's commission for seizing pirates and making reprisals, England being at war with France. Just why it was, nobody has explained, but Captain Kidd spent his time in loitering around the coast of Africa, where no pirates were to be found, until he grew quite disheartened, and, fearing to be dismissed by his employers and to be "mark'd out for an unlucky man," he started a little pirate business for himself, in which he gained more of a certain kind of fame than any of the rest; for popular tradition supposes him to have hoarded his booty and buried it. "Captain Kidd's treasure" has been sought for until the whole eastern coast of the United States is honeycombed with diggings for it; but probably he had eaten and drunk it up before 1701, when he was captured and executed in

"In revel and carousing
We gave the New Year housing,

With wreckage for our firing,
And rum to heart's desiring."

England. About this time, however, and without his valuable aid, the combined naval forces of all the nations interested in the commerce of the New World broke the power of the buccaneers, and their depredations ceased. Their story is one of the wildest, most romantic, and most terrible in the history of the world.

The trade of piracy was carried on during the eighteenth century in the region of the West Indies by unorganized bands of desperados who had all the faults and none of the greatness of the men they succeeded, and who received little attention from the world at large. At the beginning of the nineteenth century the Barataria pirates came into notice on the coast of Louisiana, taking the place of the buccaneers, but in a much smaller way. Their leaders, Pierre and John Lafitte, carried on business quite openly in New Orleans; and their settlements on the marshy islands along the coast, and their "temple," to which persons came out from the city to buy goods, were open secrets. But in the War of 1812, although the British tried to buy their services, they redeemed themselves by standing true to the American government, which had just been trying to exterminate them, and so they won public pardon and an added glamour of romance.

For the same reasons as those in the case of other island systems, the East Indies have always been infested with pirates, whose light, swift vessels run in and out of the intricate channels among the dangerous coral reefs, where government cruisers dare not follow, while the people on shore sympathize more with the pirates than with the police.

The East Indian sea-robbers are, as a rule, natives of that region — Malays, Borneans, Dyaks, and Chinese, with many half-savages of the South Sea Islands. This is more like a continuance of savage resistance to civilization than real piracy, since the pirates of the Atlantic are civilized sailors in mutiny against their own people and national commerce. The result is just as bad, however; for these East Indians are as bloodthirsty and cruel as the others, and if they do not kill their victims, or save them for some cannibal feast (as would probably happen in the New Hebrides and some other islands), they condemn them to a life of misery. But in these days of improved sea-craft, piracy, even in Malayan waters, is weak. Our consuls and government agents watch suspicious vessels; our telegraph warns the naval authorities in a moment; our steam-cruisers outspeed the swiftest craft of the black flag; our rifled guns silence their cheap artillery; and our coast surveys furnish maps so accurate that the pirate no longer holds the secret of channels and harbors where he can safely retreat. If, therefore, the old "Redbeards" should come back to life and try to be kings of the sea, as they rejoiced to be a couple of centuries ago, their pride would soon be humbled, and they would gladly return to their graves and their ancient glory.

There is a form of sea-roving which has been at times not very different from piracy; it is called *privateering*, and history shows a good many cases where it has degenerated into sea-robbery pure and simple.

A privateer is a ship, owned by a private citizen or citizens, to which authority is given by a government to act as an independent war-vessel. Its commission is called a "letter of marque" (*lettre de marque* in French), entitling it to "take, burn, and destroy" a certain enemy's property on the sea or in its ports. It has no right, of course, to attack any one else.

The object and plea of the government issuing commissions to privateers is that thus a great many more armed vessels can be sent afloat than the government has money to equip, and that consequently far more damage

MALAY PIRATES ATTACKING A STEAMER.

will be done to the enemy, by crippling his trade and resources, than regular men-of-war alone can accomplish. Private capital has been willing to take the risk because rewarded by a large share of the prizes; and from the end of the fifteenth to the middle of the eighteenth century this was one of the most profitable of marine industries, for then nearly universal wars made almost any capture legitimate. In the earlier times even the limited regulation that came later was absent, and there was small choice between a privateer and a pirate. Queen Elizabeth found the hundreds of privateers which she had commissioned against the Spanish and Dutch preying upon her own people, and robbing fishermen, coasters, and small shore towns, to such an extent that she had to suppress them as bandits. Those were the times when Hawkins could use a royal fleet to wage war upon the Spanish colonies for private reasons; and when his ablest lieutenant, Drake, could make his notable journey around the world a history of robbery and slaughter. On the west coast of South America he spent months in destroying Spanish vessels and ravaging and burning settlements; yet it was thought remarkable, when he returned from his circumnavigation of the globe, that the Queen hesitated somewhat before recognizing his great achievements as a seaman, for fear of complications with Spain!

Spain, in those days of first harvest from her American possessions and

the East Indies, was the prey of everybody on the high seas able to rob
her, and formalities were joyously disregarded by both sides.   Her galleons
carried precious cargoes of spices, silks, and East India goods around the
Cape, and brought silver ingots and gold bars from the Spanish Main.
They were usually convoyed by regular war-ships, and had to run the gant-
let of the enemy's fleets whenever Spain happened to be openly at war with
somebody, as was usually the case; and otherwise must escape buccaneers
in West Indian waters, Malayan and Chinese pirates in the far East, and
irregular sea-rovers along the West African coast, while the corsairs made
the Mediterranean route doubly dangerous.

The gradual growth of organized navies, the development of international law, and the increasing organization of the civilized world gener-

PAUL JONES' FIGHT IN THE "BON HOMME RICHARD" WITH THE "SERAPIS."

ally, slowly tamed these wild practices and reduced privateering to some
sort of control.   Thus Jean Bart, the popular hero of French naval history,
who flourished toward the end of the seventeenth century, was recognized
and supported by the French monarch as a free-lance in the Mediterranean,

because his humble birth prohibited him from taking a commission in the
regular navy, which amounted to a sort of apology for his deeds.

During the wars of the United States with England privateering was
extensively practised on both sides, and was of especial value to the Americans. Congress issued private commissions as early as March, 1776, and
the ablest statesmen upheld it as a means of employing the ships, capital,
and thousands of seamen that must lie idle when the enemy's cruisers were
ranging the ocean highways unless permitted to arm themselves and assist
the government in an irregular warfare, trusting to the value of their captures for remuneration. That the chance of such reward was enough induce-
ment is shown by the fact that during the first year of the Revolution nearly
three hundred and fifty British vessels were captured, chiefly West Indiamen, worth, with their cargoes, five million dollars. As Great Britain did
not recognize the flag of the United States, not only these, but even our
regular naval officers, were regarded by them as pirates, rather than true
privateers — Paul Jones first of all; but she never acted on this theory with
the severity that would have been visited upon true pirates.

In the naval warfare that came later between the United States and
France, privateering again flourished, and was a source of immense profit to
the principal seaports whence these swift, effective Yankee vessels were
despatched. No less than three hundred and sixty-five American privateers
were sent out between 1789 and 1799, and swept the seas almost clean of
the French merchant flag.

Then came the second war with Great Britain, which was fought over
a question of the sea rather than of the land,—the right of search
claimed by the British,—and once more American and British privateers
swarmed upon the highways of commerce. Of our merchant ships in all
parts of the world, about five hundred were lost; but this was more than
paid for, since our two hundred and fifty privateers captured or destroyed,
during the three years and nine months of the conflict, no less than sixteen
hundred British merchant vessels of all classes.

This disparity of results was largely due to the greater number of English merchant vessels, but is also to be credited to the superior speed and
handiness of the Yankee vessels, most of which were " Baltimore clippers,"
topsail-rigged schooners with raking masts, that could outsail and out-
manœuver anything afloat. "They usually carried from six to ten guns,
with a single long one, which was called 'Long Tom,' mounted on a swivel
in the center. They were usually manned with fifty persons besides officers,
all armed with muskets, cutlasses, and boarding-pikes."

An English writer, Mr. R. C. Leslie, is of the opinion that this type of vessel grew out of models in vogue in the West Indies, long before, for the small piratical craft that made those waters the terror of travelers.

These Baltimore clippers, too, enlarged and square-rigged, but still the fastest things on the western ocean, formed the craft with which the slave-trade was continued between Africa and America long after it had been condemned by the civilized world. For many years previous to the American Civil War, which put an end to the larger part of the traffic by destroy-

UNITED STATES FRIGATE "CONSTELLATION" OVERHAULING THE SLAVER "CORA."

ing its market, England and the United States kept squadrons patrolling the African coast to arrest the slavers and free their "cargoes."

What wild, wild tales of the sea do these reminiscences of piracy, privateering, and the slave-chase bring to mind—tales of horror, and yet full of such deeds of daring and romance and fierce delight as must stir the heart in spite of brain and conscience!

Pirates are things of the past — no more to be feared except in a small way in the Malayan and Chinese archipelagoes. The African slave-trade is extinct, so far as shipment across the ocean is concerned, save where, now and then, an Arab dhow steals with its black cargo along the East African foreland, or flits across the Gulf of Aden or the Red Sea. Privateering has been forbidden by international treaty among the larger

European powers, which now recognize that trade goods, even of belligerents, must be held safe in the ships of neutrals (except articles declared contraband of war), because the business of the world cannot stop, or even be put in jeopardy, by a quarrel between two nations. Privateering, therefore, has been abandoned in Europe as a method of war since the treaty of Paris in 1856, though Prussia came pretty near it in 1870, by organizing what she called a volunteer fleet, and Spain reserves the privilege of commissioning privateers.

The United States, however, and some other countries whose policy or ability forbid them to have a large navy, would not enter into the European agreement above mentioned, mutually to abstain from privateering, on the plea that to do so would be to yield the most powerful weapon of a nation weak in naval armament and sea commerce, against any of many possible enemies whose large sea-borne commerce would expose it to the most serious wounds. In our Civil War the President issued no letters of marque, although authorized to do so. It was customary to speak of the Confederate cruisers *Alabama, Shenandoah, Florida,* etc., as privateers, or even pirates, and they actually played the part with a success woeful to us of the North, and to Great Britain, which had to pay for the damages caused by the *Alabama ;* but, strictly speaking, they were neither, because commissioned by a temporary but regular government, whose flag might have been recognized if its arms had succeeded.

More lately (1898) the United States has announced it as its policy to refrain from privateering, though no formal signature has been given to any international agreement to that effect.

STEAM YACHT.　　"HALCYON."　　SANDY HOOK LIGHT-SHIP.　　"COLUMBIA."　　"MAYFLOWER."

A SPIN OUT TO SEA.—WELL-KNOWN YACHTS ROUNDING THE SANDY HOOK LIGHT-SHIP.

(From the painting by J. O. Davidson, owned by F. A. Hammond, Esq.)

CHAPTER IX

YACHTING AND PLEASURE-BOATING

 ACHT is a word derived from the Dutch language, which has given to the English so many of its sea-terms, meaning, originally, a fast boat, such as was built for chasing pirates and smugglers, and, later, a pleasure-boat. The latter meaning alone is now kept in view by the word, which is properly applied to anything designed and used for pleasure-sailing, whether moved by sails, steam, or electricity.

In Great Britain, where yachting, as we now understand it, arose, it was not until about 1650 that races between pleasure craft began to be sailed on the Thames and in the quiet waters about the Isle of Wight, while the first yacht-club was not formed until 1720 (at Cork, in Ireland). Even then, a century elapsed before yachting as a sport attracted much attention even among the British, famous for their love of the sea. In 1812 a "yacht-club" was founded at Cowes, in the Isle of Wight. It received a new impetus and became the "Royal Yacht-Club" in 1817, the Prince Regent having joined it, and in 1833 was again reorganized by King William III as the "Royal Yacht Squadron," the designation it bears to-day. It carried on races, or regattas, as they soon came to be called (borrowing from the Italians a term descriptive of the old Venetian gondola races), but all sorts of cruising-boats were matched against one another, classified by a tonnage rule with no allowances for size or any of the systems by which contestants are now classified and equalized.

By this time, however, there was peace on the North Atlantic, and many a good seaman was free to turn his attention to enjoying and improving the tools of his profession. By this time, also, the Americans had made great headway as ship-builders and seamen, and by rivalry with the Old World for trade, and by experience in the Newfoundland fisheries and the West Indies fruit-trade, had acquired a skill in building and rigging ships that astonished the world by their speed and weatherly qualities. It was natural

that these ideas should influence pleasure craft on this side of the water, as
Great Britain's long sea-struggles had influenced its sailors; and when, in
1844, the New York Yacht-Club was founded, the conditions were favorable for beginning that home development of yachting as a sport which was

"AMERICA" (AS ORIGINALLY RIGGED) AND "MARIA."

soon to place the Americans and Canadians among the leading yachting
peoples of the world, and to lead to those international tests of speed that
nowadays excite so wide-spread and intense an interest.

The great preponderance in numbers and value of pleasure-vessels in
the United States, and in the number of clubs and club-members, is due not
only to our large population and long coast-line, but to the great extent of
inland waters furnished by our rivers and interior lakes, and to the prevalence of bays or protected lagoons, such as Narragansett Bay, the Great
South Bay of Long Island, New York harbor, Delaware and Chesapeake
bays, and the long series of "sounds" that border the southern Atlantic coast
from Barnegat to Biscayne. The Great Lakes are bordered by yacht-clubs
on both sides, and furnish space and weather for quite as serious work as
tries the skill of ocean navigators, while a hundred smaller lakes make fine
pleasure-waters and excellent training-grounds for fresh-water sailors.

Though the first regatta in America was sailed in 1845, little over half a century ago, the evolution of American yachts began with the building of the sloop *Maria* by Robert L. Stevens, one of that family of remarkable inventors, who had already devised the first practical screw-steamer, and afterward created the *Monitor*. Her model, as we learn from an excellent article in "The Century" for July, 1882, by S. G. W. Benjamin, was suggested by the low, broad, almost flat-bottomed sloops employed to steal over the shallows of the Hudson and the Sound — vessels depending upon beam rather than on ballast for stability, and imitated by many of our coasters, which are so stiff that they sometimes make outside voyages without either cargo or ballast; but the *Maria* had a long, sharp, hollowed bow, whence she expanded aft, with little taper at the stern, so that her deck-plan was that of an elongated flat-iron. The principal novelty about her, however, was the use of two "center-boards."

A center-board is a plate of wood or metal, suspended, usually by a corner pivot, within a sheath or box in the waist, which can be let down through the keel into the water, so as to form an adjustable keel. It is the most convenient form of a very old device for preventing a boat's drift to leeward, or tendency to capsize under the pressure of the wind. In earliest times, a mat was hung over the side. Later this was replaced by the leeboard, apparently a Dutch invention, which may still be seen on the canal barges in Holland, and which was a feature of the pirogues or periaugers (shallow double-ended sailing-canoes) that in early times formed almost the only type of small sail-boat in New York waters. Two other novel, foreshadowing features possessed by Mr. Stevens' boat, were the use of rubber compressors on the traveler of the main boom to ease the strain of the sheet (rubber is applied in many places about modern rigging), and the bolting of lead to the keel as outside ballast.

The *Maria* justified the expectations aroused by these and other novelties in hull and rig by beating everything in existence, until a Swedish gentleman in New York constructed a much smaller boat, the *Coquette*, on very different lines, for although only sixty-six feet long she drew ten feet of water; and in a match on the open sea she beat the *Maria* easily, showing the superiority of the deep-keeled model for windy weather.

Profiting by these experiences and widely gathered information, a new designer essayed the task of making a still better yacht. This was George Steers, the son of a British naval captain and ship-modeler, who had become an American naval officer and was the first man to take charge of the Washington navy yard. He built several graceful and fleet-winged sloops, famous in their day, such as the *Julia*, David Carl's *Gracie*, and many pilot-boats and ships. His most celebrated production, however, and the one which gave our yachtsmen an international reputation and established their method of pursuing recreation as the foremost American sport, was the *America*, from which the "America Cup" races take origin and name.

The origin was really accidental. When the first World's Fair was to be held at the Crystal Palace in London, one of the attendant festivities was a great national gathering of British yachts in their favorite harbor, Cowes, at which, it was announced, foreign yachtsmen were to be welcome, especially Americans. In preparation for it, John C. Stevens, of Hoboken, then Commodore of the New York Yacht-Club, and some of his friends, ordered a new yacht from George Steers with which to cross the Atlantic and meet the English racers. This new boat, completed in the spring of 1851, and named *America*, was schooner-rigged, but had raking masts, no topsails except a small main gaff, and only one jib, whose foot was laced to a boom. Such was the style of the day; but later she was changed in rig so as to carry far more and bigger sails, more like those of a modern schooner-yacht.

The moment she arrived in Cowes, in the early summer of 1851, her superiority in speed was conceded, and no British captain would consent to meet her; but finally a match was extemporized, open to all nations, for which a prize was offered in the form of a cup presented by the Royal Yacht Squadron — not by the Queen, as usually said. Fifteen yachts responded, but none showed what it could do, for there was little wind, and the cup was awarded to the *America* more in general acknowledgment of its excellence than because of any great performance there. Not much importance was attached to the incident, but the silver tankard was brought home and left to ornament Commodore Stevens' drawing-room until 1857, when its owners dedicated it to the purpose of a perpetual challenge cup, in charge of the New York Yacht-Club, for international races under specified conditions. Fifteen years elapsed, however, before the first contestant appeared.

The *America* had differed prominently in shape from all her opponents at Cowes, by having fine hollowed bows and a wide stern, instead of the bluff bows and narrowing after part — the "cod's head and mackerel's tail" pattern — of English craft; she also had sails that hung very flat instead of bellying out under the wind as was the foreign style. In these directions British yachtsmen saw good, and tried to improve; but they would have nothing to do with center-boards, and clung to their cutter-rig. We, on the other hand, had gained ideas as to improving rig, especially in the schooners, and in the bestowal of ballast, outside and in.

At length, in 1870, an English schooner, the *Cambria*, came over to compete for the cup, and was pitted against a fleet of crack yachts off Sandy Hook; but again the wind was so light that the boats did little more than drift. The Englishman, nevertheless, was outdrifted by nine others, and the leader was the little sloop *Magic*, which became the custodian of the cup. The next year, however, another challenge was received, and the British keel-yacht *Livonia* appeared and was defeated by the American keel-schooner *Sappho*, which, under a new rule, had won her right to defend the cup by first beating in preparatory ocean races all other rivals for the honor. As this contest was between single representative yachts, tried in

"GENESTA," "IARA," AND "IREX"—THE BRITISH TYPE OF CUTTER OF 1884-85.

"Galatea," 1885, belonged to the same type.

five races, and in all sorts of weather, it was a fair and conclusive measure of comparative qualities. The next yacht to come after the international cup was the Canadian *Countess of Dufferin*, which was promptly defeated by the *Magic* in 1876. Five years later another Canadian appeared, the *Atalanta*, differing from previous contestants in being a single-masted center-

board yacht: but her rigging and finish were so bad that her excellent
model could not save her from defeat (1881) at the hands of the elegant
iron sloop *Mischief*, which had been built especially for the race, and had
won her foremost place through severe trial races, as before.

Up to this time, as Mr. W. P. Stephens tells us in "The Century" for
August, 1893, whence many of the portraits of these racers have been taken,
no pleasure-boats had been built except after the rule of thumb — some
practical sailor whittled out a model according to his ideas, and the builder
followed it.

Systematic designing was unknown, and . . . one type of yacht was in general use, the
wide, shoal center-board craft, with high trunk cabin, large open cockpit, ballast all inside (and
of iron, or even slag and stone), and a heavy and clumsy wooden construction. Faulty in every
way as this type has since been proved, in the absence of any different standard it was considered
perfect, and open doubts were expressed of the patriotism if not the sanity of the few American
yachtsmen who, about 1877, called into question the merits of the American center-board sloop,
and pointed out the opposing qualities of the British cutter — her non-capsizability, due to the
use of lead ballast outside of the hull; her speed in rough water; and the superiority of her rig
both in proportions and in mechanical details.

A wordy warfare over these types raged for several years, gaining strength with the building
of the first true English cutter, the *Muriel*, in New York in 1878, and bearing good fruit a year
later in the launching of the *Mischief*, an American center-board sloop, but modified in accordance with the new theories. The plumb stem, the straight sheer, and higher free-board, with
quite a shapely though short overhang, suggested the hull of the cutter, and though quite wide
— nearly twenty feet on sixty-one feet water-line — she drew nearly six feet. Even with her
sloop rig she was a marked departure from the older boats of her class, especially as she was
built of iron in place of wood, and consequently carried her ballast, all lead, at a very low point.

One of the results of this controversy was the sending to this country,
from Scotland, of a little ten-ton racing cutter, the *Madge*, purely to show
what capabilities lay in "a deep, narrow, lead-keeled craft with the typical
cutter rig." The only American able to beat her was the *Shadow*, a famous
Herreshoff sloop of unusual depth, and she did it but once. Nevertheless,
the controversy was not decided in the United States, and the Britishers
thought it worth while to try to give us another lesson. In 1884 they
launched two big cutters, *Irex* and *Genesta*, and in 1885 a third, *Galatea*;
and Sir Richard Sutton, owner of *Genesta*, and Lieutenant William Henn,
R. N., owner of *Galatea*, challenged for the America Cup.

Then the question arose: What should be done to meet them? The
British cutters differed from those previously met, in that they were built
for racing, not for general use — were "racing machines" instead of cruising-yachts. To meet this, a scientific designer of marine vessels, Mr. A. Cary
Smith of New York, was called upon to produce a moderately deep, center-

board, iron sloop-yacht on the lines of the *Mischief*, but much larger, and he produced the *Priscilla*. But while she was building there was quietly begun another yacht, the *Puritan*, owned and built in Boston from designs by an almost unheard-of architect, Mr. Edward Burgess, who previously to this performance had been renowned only as a student of insects!

"The stout oak keel of the new *Puritan* was laid upon a lead keel of twenty-seven tons, carried down into a deep projecting keel; the plumb stem, the sheer, and the long counter suggested the British cutter rather than the American sloop; the draft of eight feet six inches was greatly in excess of all of the old center-board boats, and the rig was essentially that of the cutter rather than of the sloop."

A struggle decided that she was better than the *Priscilla*, and in the cup races in September she proved herself better than the famous English cutter *Genesta*.

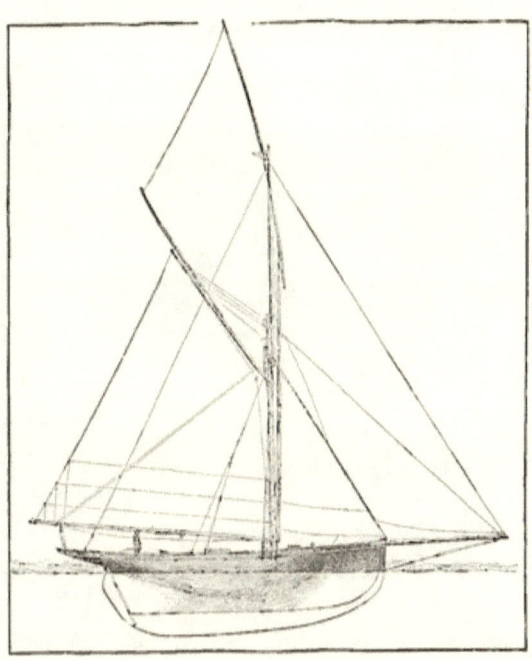

THE CUTTER "MURIEL," SHOWING THE ENGLISH DEEP-DRAFT TYPE OF BUILD AND RIG.

Nevertheless, when the *Galatea*, whose challenge had been postponed until 1886, came out, the *Puritan* had already been distanced by an American rival, the *Mayflower*, practically a larger copy of herself, as *Galatea* was of *Genesta*, and, therefore, a lead-keeled center-board boat, having a cutter-like rig. Trial races showed that the *Mayflower* was able to beat all her beautiful predecessors, and again the British contestant was obliged to take a defeat and leave the prize in New York.

The result of this last contest (1886) was to cause British yachtsmen to abandon their old tonnage rule of measurement and adopt the far better modern one of load-line and sail-area measurement. Another challenge

13

immediately came from Glasgow, supported by a boat named *Thistle*, built under the new rule; and to oppose it Mr. Burgess built the *Volunteer*, which differed from its predecessors mainly in increased draft and tendency toward the cutter model. She easily beat the *Thistle*, and the discouraged foreigners rested for some years before trying again to wrest from us the coveted trophy. In 1891, however, there

" PURITAN."

" MAYFLOWER."

came to New York, from the yards of the Herreshoff Brothers, in Rhode Island, a new forty-six-foot yacht, which soon put the fame of the *Volunteer* and all her glorious rivals into the background. This was the *Gloriana*, "remarkable as a daring and original departure from the accepted theories." The radical novelty in her form consisted in the great cutting away of her bulk under water while preserving the full extent of the water-line, and the making of a very deep, heavily loaded keel, trusted for stability. Her hull was also novel, consisting of a double skin of thin wood on steel frames, while the upper part of the hull projected excessively at both ends. She was everywhere a winner, and was immediately followed by a smaller boat, the *Dilemma*, whose keel was an almost rectangular plate of steel, the ballast, which alone was trusted for stability, being in the form of a cigar shaped cylinder of lead bolted to the lower edge of the " fin," as this kind of keel was appropriately styled. Many boats of this pattern were soon afloat, most of them highly successful at home and abroad, and carrying a surprising spread of canvas.

The year 1893 brought another challenge for the cup in the person of Lord Dunraven, sailing the yacht *Valkyrie*, but he was met by a new, well-proved Herreshoff fin-keel, the *Vigilant* (built of a new alloy — Tobin bronze), and handsomely defeated. The following season the *Vigilant* went to England, and found herself equally overmatched by the *Britannia*, owned by the Prince of Wales, while *Valkyrie II* was wrecked. In 1895 Lord Dunraven sent a second challenge, backed by a new *Valkyrie (III)*; and this produced a fresh American contestant, again designed and built by the Herreshoffs, named *Defender*. The races came off amid intense public excitement, outside of Sandy Hook, but were most unsatisfactory; "in the first, *Defender* won; in the second, *Valkyrie* was disqualified as the result of a foul, and Lord Dunraven declined to sail a third."

Such has been the history of this long series of races for the America Cup, and such the development of its defenders; but while they and their

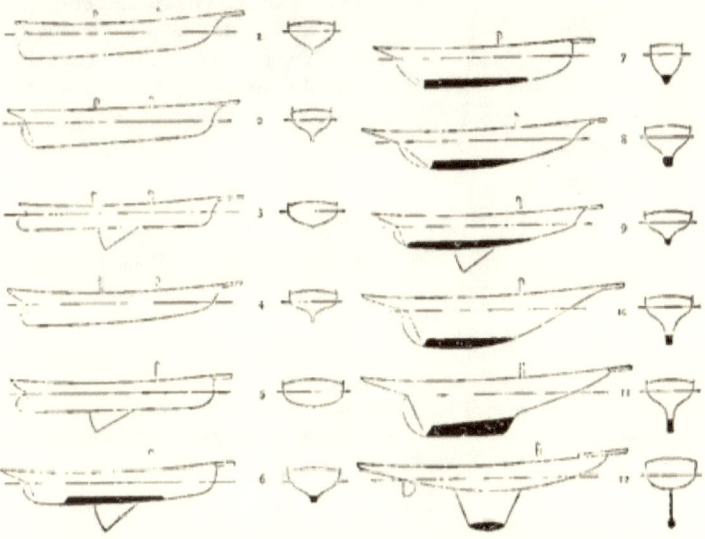

COMPARISON OF OLD AND NEW TYPES.

1. "America," 1851, water-line 90 feet. — 2. "Cambria," 1868, water-line 100 feet. — 3. "Magic," 1857-69, water-line 79 feet — 4. "Sappho," 1867, water-line 120 feet — 5. "Mischief," 1879, water-line 61 feet — 6. "Puritan," 1885, water-line 81 feet. — 7. "Genesta," 1884, water-line 81 feet. — 8. "Thistle," 1887, water-line 86 feet — 9. "Volunteer," 1887, water-line 85 feet. — 10. "Gloriana," 1891, water-line 45 feet. — 11. "Wasp," 1892, water-line 46 feet. — 12. "El Chico," 1892, water-line 25 feet.

work have stimulated interest in yachting all over the world, they have really not influenced it greatly, because all of the later boats competing were not practical yachts, in which one might cruise and live afloat, and enjoy life with his friends, but "machines" in which every quality tending to comfort and safety was sacrificed to the requirements of speed. In fact, the owners of these "big boats" kept small, handy, comfortable yachts for their own enjoyment, and the racers were as a rule sailed by a skipper and crew of professional racing sailors.

There are said to be over two hundred yacht-clubs in the United States, enrolling about four thousand yachts, an eighth of which are steam or elec-

tric boats, scattered wherever any water suitable for the sport exists. With the lakes and rivers we have nothing to do, except to say that the yachtsmen of Montreal and Quebec are really salt-water sailors, for they cruise in the Gulf of St. Lawrence and elsewhere at sea as well as their fellow-sportsmen of New Brunswick and Nova Scotia. At the other extreme the Havana Yacht-Club has American members who take their boats to the West Indies every winter. Bermuda is another fa- vorite resort, and the scene of lively races with a local, narrow sort of craft, called a "flyer," which will beat almost any- thing if only it can be kept right side up.

THE NEWPORT CATBOAT.

On the Pacific coast, . . . wherever there is a bay that will afford a harbor, and a town that will support people, the yacht is used as a vehicle of pleasure. . . . Many of the San Fran- cisco boats are large schooners, a number are powerful sea-going sloops, while of smaller craft there is an abundance of almost every type, although the New York catboat and the flat-bot- tomed sharpie of Long Island Sound are seldom met with, and seem not to be in favor. . . . Pacific yachters appreciate the good points of the yawl, for the squalls which blow over the waters of the west coast are sudden and severe, and no rig meets these conditions of weather so well as does the yawl.

The most important and numerous yachting interest of the country, how- ever, as would be expected, is along the northeastern seaboard, where, measured by numbers and the investment in boats, wharves, club-houses, and equipments generally, it surpasses any other district in the world. More than one hundred clubs exist between Maine and Philadelphia.

The earliest form of yacht [as Mr. F. W. Pangborn reminds us in "The Century" for May, 1892] was, of course, a rowboat with a sail. . . . From the primitive sprit-sail pleasure-boat comes the ever-present and universally favored center-board catboat, a type of yacht which, for

RIG OF THE YAWL.

speed, handiness, and unsafeness, has never been surpassed. Keel catboats are also built, but the typical American "cat" is the center-board boat of light draft, big beam, and huge sail. The two objectionable points about boats of this class are their capsizability, and their bad habit of yawing when sailing before the wind. Yet the cat is the handiest light-weather boat made. It is very fast, quick in stays, and simple in rig; but it can never become a first-class seaworthy type of yacht. It belongs among the fair-weather pleasure-boats. . . .

From the center-board catboat grew the jib-and-mainsail sloop, a type of yacht which has always been noted for its great speed and general unhandiness. Small yachts of this kind are always racers, and the interest in racing is sufficient to keep them in the lists of popular boats. In design they are like the catboats, the only difference being in their rig. These two boats, the center-board cat and the jib-and-mainsail sloop, are what yachters call "sandbaggers"; that is to say, their ballast consists of bags of sand which are shifted to windward with every tack and thus serve to keep the yachts right side up. A boat ballasted in this manner can carry more sail than rightly belongs on her sticks, but she cannot be very safe or comfortable. Her place is in the regatta. It is not beyond the truth to assert that the sandbaggers constitute probably two fifths of the total of small yachts. They will never cease to be popular, for the reason that speed and sport are synonymous terms with a great many yachters, and no one can deny that these boats, like Brother Jasper's sun, "do move."

Passing the sandbaggers, the next popular and most universally used yacht is the ballasted sloop. A sloop may be a center-board boat, or a keel boat, or a combination of both. She has only one mast and carries a topmast. Her sails are many, and, like the cutter, she is permitted to carry clouds of canvas in a race. Technically speaking, a cutter differs from a sloop only in one point, as the terms "sloop" and "cutter" really apply to the rig of the yacht. The cutter has a sail set from her stem to her masthead; the sloop has not. This sail is called a forestay-sail, and its presence marks the cutter-rig. The term "cutter," however, is usually applied to the long, narrow, deep-keeled vessel, and has in common parlance grown to mean a boat of that type. It is in that sense that it is generally understood. It is worthy of notice that nearly all yachters

DRAWN BY W. TABER, FROM A PHOTOGRAPH BY WALTER SCARGANON.   ENGRAVED BY A NEGRO.

A SANDBAGGER SLOOP.

13*

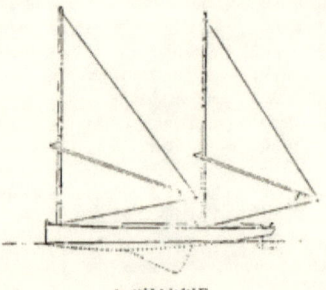

A SHARPIE.

who cruise about in summer, and especially those who are fond of speedy boats, use either sloops or cutters; and it is remarkable to see how much comfort can be found in boats of these types, even when quite small. . . .

The average yachting man, if he be of that stuff of which good seamen are made, soon finds his chief delight in being master of his own vessel. He likes to feel that it is his skill, his prowess, his intellect, that rule the ship in which he sails: and finding this complete mastery of the vessel to be impossible aboard a big boat, he longs for one which he can handle alone. This independent and sportsman-like instinct of the American yachter has culminated in a liking for certain classes of very small boats,—"single-handers" they are called,—and this liking has given impetus to the building of some little vessels which are really marvels in their way. Simplicity and handiness of rig have been considered in their construction, and this has led in many cases to the adoption of what is known as the yawl style, a rig which for safety and convenience has never been surpassed by any other. The yawl is really a schooner with very small mainsail. For small cruising-yachts it is an excellent rig, and preferable to the cat rig. Cat-yawls are also in use; they are merely yawls without jibs. With such rigs as these a yachter can go alone upon the water without fear of trouble and with no need of assistance. Naturally, with men of moderate means who love the water, these small single-handers have become very popular. Some of them are not over sixteen feet long, yet the solitary skipper-crew-and-cook, all in one, of such a boat finds in his yacht comfortable

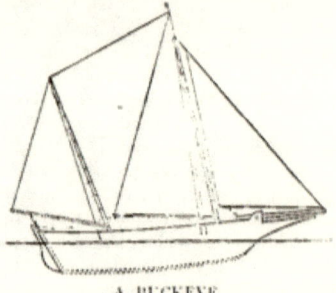

A BUCKEYE.

sleeping-quarters, cook-stove, dinner-table, and all necessary "fixings." The ingenuity displayed in fitting out the cabins of these little boats is quite remarkable.

Of the many nondescript rigs which are applied to small yachts, two are in common use. One of these is the sharpie, a simple leg-o'-mutton rig used with flat-bottomed boats. Large sharpies have been built with fine cabin accommodations, and such boats are particularly adapted to the shoal waters of the South. They are fast sailers, but, owing to their long, narrow bodies and light draft, are not always trustworthy. They are cheaper to build than boats of other designs. . . .

Buckeyes are favored only in the South. Originally the buckeye was a log hollowed out and shaped into a boat, and was used by the

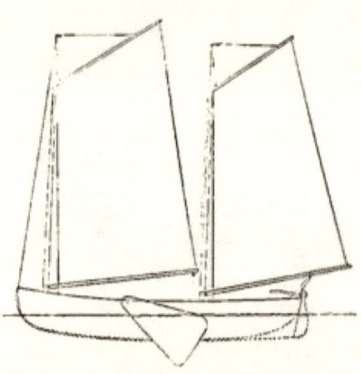

OLD-STYLE PIROGUE WITH LEEBOARD.

negroes. To-day, however, buckeyes are built upon carefully drawn plans, and many of them are excellent vessels. They are common on the coast waters south of the Delaware Bay, and are used chiefly for hunting-boats, their cheapness, handiness, and roominess rendering them useful to the sportsman. A true buckeye is a double-ender, but some large ones have been built with an overhang stern, which destroys the ideal and creates a new kind of craft.

A few years ago the sailing public was surprised by the appearance upon the waters of a spider-like contrivance which its friends said was a "catamaran." This new claimant for yachting favor was like the raft of the South Sea Islanders only in name; in fact, it was not a catamaran at all, but a new device for racing over the water by means of sails. Wonderful feats were predicted for the future of the catamaran, and it certainly did accomplish something; but after a long and fair trial (for the yachter, no matter how bigoted he may be, will always try a new boat) it was discarded as a useless, dangerous, and decidedly unsatisfactory kind of craft. . . .

Leaving the discussion of the odds and ends of yacht styles, we come, by natural progress, to a type which is destined to greater popularity as time goes on, and yachters learn the ways of the sea and the best methods of dealing with them. Although the schooner is generally deemed a big yacht, it is nevertheless a fact that small schooners are desirable boats to have, and that the number of schooners of small tonnage is increasing. There is no denying the advantage of the schooner's rig over that of the sloop. A schooner of forty feet is handier, safer, and less expensive to run than a forty-foot sloop. The rig of the schooner is peculiarly adapted to all weathers, and a small crew can handle such a vessel with ease, when to manage a sloop of equal size would require the best efforts of "all hands and the cook." The reason for this is that the schooner's sails can be attended to one at a time, which is not the case with the big-mainsail sloop.

It is the small yachter [Mr. Pangborn declares in conclusion] who gives to the sport its wide popularity, and makes yachting so universally loved by men who are fond of aquatic pleasuring. The small yachter is everywhere upon the waters. From the coast of Maine, from the shores of the harbor of the Golden Gate, from the beaches of the Atlantic seaboard, and from the borders of the inland lakes, he can be seen, all summer long, sailing about in his little vessel, and enjoying in all its fullness the excitement and delight of this most noble and health-giving sport. With a pluck and energy that mark the true lover of the sea, and a tact and skill that bespeak the real sailor, he handles his little craft, in fair weather and in foul, in a manner that leaves no room for doubt as to its fitness for the work which he is doing; for, whether he sail alone, or with the help of his friends, or that of a hired man to run his boat, he is always the master of his vessel,—which is seldom the case with the proprietor of the big boat,—and is in reality a "yachtsman" under all circumstances, at all times, and in all weathers. He must be cool-headed and calm in times of peril, affable and courteous on all social occasions, and generous and prompt to respond to all calls upon his courage — in brief, a gentleman.

FROM A PHOTOGRAPH BY WALTER BLACKBURN.

YACHTS WAITING FOR A BREEZE.

THE "ADLER" PLUNGING TOWARD THE REEF AT SAMOA.

CHAPTER X

DANGERS OF THE DEEP

 EITHER ships of the stanchest steel, nor seamen however skilful, nor pilots never so knowing, can wholly avoid the dangers of a seafaring life. Experience in reading the signs of the ocean and of the skies, surveyors' charts of coasts and harbors, added to the appliances of powerful modern machinery, have lessened the perils, it is true, since the old times; yet even now ships sail proudly out of sunny havens, their topsails watched by loving eyes till they disappear at sunset, and are never seen again. On a calm day in 1782 the great hundred-gun line-of-battle-ship *Royal George* sank at her anchors in the harbor of Spithead, carrying down almost a thousand souls; thirty years ago the *Captain*, then one of the finest of England's steam turret-ships, capsized at sea, and not a man survived. Each of these vessels was perhaps the best of its kind in the world. No better navigators exist than naval officers, yet they ran the historic old steam-frigate *Kearsarge* on Roncador Reef, in the Caribbean Sea, in broad daylight, and left her there a total wreck. Not a year passes that does not record some dire calamity on the ocean, and many lesser accidents.

The wild oceanic storms are responsible for fewer of these than anything else — I mean the mere power of wind and waves in the open sea. When a captain has sea-room, and knows in advance, as he almost always may, of the coming of a storm, so that he can make everything snug, the loss of his vessel, or even serious damage to her, is not common. Yet the mere violence of the gale has overturned, beaten down, and extinguished the greater part of the Newfoundland fishing-fleet again and again, and doubtless many of the ships that are recorded as "missing" have been sunk simply by overwhelming waves.

Certain rare and extraordinary mishaps nevertheless may meet a vessel in the open ocean. One of these is a stroke of lightning, powerful enough to set a ship on fire in spite of her lightning-rods, and such a fire is likely

never to be quenched.   Another extraordinary occurrence would be an overwhelming waterspout, such as not infrequently is seen in the tropics, especially along the Chinese coast, where it often plays havoc with fishing junks.
A third unusual, yet possible, peril is the meeting with those waves of
sudden and extraordinary size and volume which sometimes engulf vessels in
storms that otherwise might be safely weathered, or are surmounted only by
a miracle, as it were.   These are said to be produced in some cyclones, as
one of the effects of that whirling form of storm, and are often called tidal
waves, but the tide has nothing to do with their formation or progress.

THE U. S. S. "ONEIDA" AFTER COLLISION WITH A STEAMSHIP.

To say that a ship in mid-ocean might be destroyed by an earthquake
seems paradoxical and absurd, yet it is true.   Whenever a subterranean
convulsion occurs beneath or at the edge of the sea, the water will be agi-
tated in proportion to its force.   Strike a tub of water a gentle tap and see
how its liquid contents shiver and ripple.   Watch a railway train running
at the edge of a body of water, and observe how the water trembles under
the percussion of the wheels upon the ground.
Earthquake shocks give rise sometimes to great disturbances, either by
a direct jar to the water, or by setting in motion waves whose rolling does

damage, especially in confined harbors. Sometimes a port will be suddenly invaded by a wave, the cause of which was an earthquake, which rolls in upreared like a wall, and carries death and destruction in its course. The principal port of the island of St. Thomas, in the West Indies, was once devastated by this means. The incoming wave is said to have been over forty feet high, and broke inland, destroying much property and causing many deaths. "So tremendous was this breaker that it landed a large vessel on a hillside half a mile from the harbor."

Such catastrophes are not uncommon in volcanic districts, where the ocean retorts with terrible vengeance when it is struck by the land. That appalling explosion in 1883 of Krakatoa, in the Strait of Sunda, was followed on neighboring coasts by a series of vast billows that rolled inland, deluging a wide extent of shore, sweeping away over 150 villages, and crushing or drowning more than 30,000 persons. Within a few years the coasts of northern Japan have been inundated repeatedly by earthquake waves with similar dire calamities, and they are likely to occur again. Now and then earthquakes are felt even in the open sea, far from land. Thus, Captain Lecky, a scientific writer upon the sea, tells us that in one instance where he was present, the inkstand upon the captain's table was jerked upward against the ceiling, where it left an unmistakable record of the occurrence; and yet this vessel was steaming along in smooth water, many hundreds of fathoms deep. "The concussions," he says, "were so smart that passengers were shaken off their seats, and, of course, thought that the vessel had run ashore." All this disturbance was, nevertheless, only the result of a shock at the bottom; and when the non-elastic nature of water is considered, the severity of the jar is not surprising.

It would seem as though in the vast breadth of the "world of waters," and with nothing to obstruct the view, two ships might easily give one another a wide berth; yet a collision is one of the ever-present dangers of voyaging, even far from land. It is to avoid this peril that all the maritime nations have agreed upon certain signals, and "rules of the road" which are the same in all parts of the world, and without which it would now be almost impossible to carry on commerce or travel on the water.

The rules of the road say that when two vessels are approaching one another, head on, each shall turn off to the right far enough to avoid the other; that when two vessels are crossing one another's courses, the one which has the other on her starboard (right hand) must turn to starboard (the right), and go behind the other vessel, while the latter continues along her course; and that a steam vessel must always get out of the way of a sailing vessel, one at anchor or disabled, or a vessel with another in tow.

It is presumed that every ship will keep a sharp lookout, and that in the daytime two approaching ships will see each other in time to keep safely apart; but in the darkness of night none could be safe unless all carried lights by which the position and character of each could be determined.

In ancient times this matter of lights at sea was a much more troublesome one than now. We know that the Roman navy managed it somehow, and had methods of signaling by lanterns and torches. In medieval and early times, say up to a couple of centuries ago, a ship's lights were a much more conspicuous and bothersome part of her than now, when, indeed, electricity has simplified as well as perfected signaling as much as it has benefited general illumination on ship's board. In such ships as those of the Armada, and long afterward, three huge lanterns made of ornamental iron-work, sometimes large enough to enable a man to move about inside them, surmounted the elevated after-quarter; and these were filled with dozens of great candles. How important candles were in the stores of one of these old ships is shown by the fact that we still call a merchant who outfits vessels a *ship-chandler*. Regular rules were formulated for judging of a ship's position and movements, and how you ought to steer by the way these beacons grouped themselves. The introduction of whale oil gradually superseded candles and as the sperm-lamp did not require a glass house, smaller lanterns took the place of the big ones, until finally, by aid of lenses, reflectors, and kerosene, and still more lately by the use of electricity, ship's lights have become the small, handy, and powerful ones they are to-day.

The present rules as to lights are these — using the language of a United States navy officer, Lieut. John M. Ellicott, who has written many instructive and entertaining essays on sea-affairs:

When you face toward a ship's bow the side at your right hand is called the starboard side, and the side at your left hand is called the port side. On her starboard side a ship carries at night a green light, and it is so shut in by the two sides of a box that it cannot be seen from the port side or from behind. On her port side she carries a red light, and it is so shut in that it cannot be seen from the port side or from behind. If the ship is a steamship she carries a big white light at her foremast-head, but if she is a sailing vessel she does not. This white masthead light can be seen from all around except from behind. . . .

It is for the red and green lights, commonly known as the side lights, that the officer of the deck most intently watches (when the lookout warns him that lights are in sight), for by them he can tell which way the vessel is going. If her red light shows, he knows that her port side is toward him and she is crossing to his left; if it is her green light, her starboard side is toward him and she is crossing to his right; but if both the red and green are showing, she is heading straight in his direction. . . . If a vessel has another vessel in tow, she carries two masthead lights instead of one; and when a vessel is at anchor she has no side lights or masthead light, but a single white light made fast to a stay where it can be seen from all around her.

In rivers and crowded harbors it is often impossible to follow the rules of the road; and sometimes even at sea the officer of the deck of one vessel discovers that the other is not heeding the rules. Then the steam-whistle is used to tell the other vessel what the first is doing. Thus, one whistle means "I am going to the right"; two whistles mean "I am going to the left"; and three whistles mean "I am backing"; while a series of short toots means "Look out for yourself; get out of the way!"

There is one class of vessels which is most annoying to those who direct the course of large steamers. These are small fishing-vessels. On the Grand Banks of Newfoundland, on the coast of Spain, and on the coasts of China and Japan big fleets of these little vessels are found at

ELECTRIC-LIGHT SIGNALS AT SEA: ARDOIS SYSTEM.

all times. They show no lights at night, preferring to save the expense of oil, and take their chances of being sent to the bottom; but when they see a big ship rushing down upon them, they light a torch and flare it about. Often they pay for their folly with their lives. The torch is seen too late, or not seen at all, and the great iron bow of the steamship crushes into the frail little craft, perhaps cutting her clean in two; and the unhappy fishermen sink into the foaming wake of the churning propellers, leaving not a soul to tell their wives what became of them.

Signaling with lights is principally of use to men-of-war, where, also, lanterns hung in the rigging in a particular order have a definite significance. For long-distance signaling the best system is that invented by Lieutenant Very, U. S. N. These night-signals "consist of a white, a red, and a green star, each fired into the air from a pistol, so that by firing one, two, or three of them in quick succession and in different orders, with a pause between the groups, different letters or signal numbers can be made until a sentence is complete." They can be easily read from vessels twelve miles away. For nearer work the system of the Spanish navy officer, Ardois, which consists in flashing and extinguishing, by means of a switchboard on deck, a series of red and white electric lamps in the rigging, serves very well; and close at hand a signal-man waves an incandescent electric bulb by night as he would a flag by day.

It is, however, when the land is approached that the sailor's perils become menacing. Here Old Neptune is still a match for us when he asserts himself. Nevertheless, we must go upon the restless waters, and must risk a contest with their power along the coasts, where the ocean's *line of battle* may be said to be. Therefore, every effort has always been made by men on land to be of aid to their brethren at sea by erecting beacons to guide them by night as well as by day, by marking the channels, so that hidden shoals, rocks, and obstructions may be avoided, and by contrivances to save life and property when the fury of the gale renders seamanship futile, and the noble ship is cast away in the surf thundering on some wild shore, to break up in a few hours.

THE "VERY" ROCKET-SIGNAL AT SEA.

What could be more humiliating to our pride, as well as terrifying to our hearts, than such a scene as that at Samoa, in 1889, when a whole fleet of ships, including powerful men-of-war, was wrecked while at anchor in the beautiful harbor of Apia. Of small use, then, were all their charts and lighthouses, buoys and breakwaters!

The disturbed state of affairs in Samoa caused the assemblage there, during March, 1889, of three small German men-of-war, *Adler, Olga*, and *Eber*, the British corvette *Calliope*, and the American steamships *Trenton, Vandalia* and *Nipsic*. The *Trenton*, Captain Farquhar, was one of our largest war-ships at that time, and the flagship of Rear-Admiral Kimberley; the *Vandalia*, Captain Schoonmaker, was somewhat smaller, and the *Nipsic*, Commander Mullan, was still less in size. On March 15 a hurricane demolished the whole of this fleet, except one, and ten merchant vessels besides, and caused the loss of nearly

THE "CALLIOPE" ESCAPING FROM APIA HARBOR.

one hundred and fifty lives. It is an extraordinary story, which has been fully related by Mr. John P. Dunning, from whose article in "St. Nicholas" for February, 1890, the accompanying facts and illustrations are drawn.

The harbor in which the disaster occurred is a small semicircular bay, around the inner side of which lies the town of Apia. A coral reef, visible at low water, extends in front of the harbor from the eastern to the western extremity, a distance of nearly two miles. A break in this reef, probably a quarter of a mile wide, forms a gateway to the harbor. The space within the bay where ships can lie at anchor is very small, as a shoal extends some distance out from the eastern shore, and on the other side another coral reef runs well out into the bay. The war-vessels were anchored in the deep water in front of the American consulate. The *Eber* and *Nipsic* were nearest the shore. There were ten or twelve sailing-vessels, principally small schooners lying in the shallow water west of the men-of-war. The storm was preceded by several weeks of bad weather, and on Friday, March 15, the wind increased and there was every indication of a hard blow. The war-ships made preparation for it by lowering topmasts and making all the spars secure, and steam was also raised to guard against the possibility of the anchors not holding.

The wind rose to a hurricane and was accompanied by heavy, wind-driven rain, and when toward morning it became evident that some smaller ships were already ashore, and that the war-ships were dragging their anchors in spite of every effort, the whole town was awake, and much of it down by the beach seeking what shelter it could from the sleet-like blast. This night of horror gradually lightened into dawn, when it was seen that all the war-ships had been swept from their former moorings and were bearing down toward the inner reef. The decks swarmed with men clinging to anything affording a hold. The hulls of the ships were tossing about like corks, and the decks were being deluged with water as every wave swept in from the open ocean. Several sailing-vessels had gone ashore in the western part of the bay. Those most plainly visible now were the *Eber*, *Adler*, and *Nipsic*, very close together and only a few yards from the reef.

The little gunboat *Eber* was making a desperate struggle, but her doom was certain. Suddenly she shot forward, the current bore her off to the right, and her bow struck the port quarter of the *Nipsic*, carrying away several feet of the *Nipsic's* rail and one boat. The *Eber* then fell back and fouled with the *Olga*, and after that she swung around broadside to the wind, was lifted high on the crest of a great wave and hurled with awful force upon the reef. In an instant there was not a vestige of her to be seen. Every timber must have been shattered, and half the poor creatures aboard of her crushed to death before they felt the waters closing above their heads. Hundreds of people were on the beach by this time, and the work of destruction had

"THE SAMOANS STOOD BATTLING AGAINST THE SURF, RISKING THEIR LIVES
TO SAVE THE AMERICAN SAILORS."

occurred within full view of them all. They stood for a moment appalled by the awful scene, and a cry of horror arose from the lips of every man who had seen nearly a hundred of his fellow-creatures perish in an instant. Then with one accord they all rushed to the water's edge nearest the point where the *Eber* had foundered. The natives ran into the surf far beyond the point where a white man could have lived, and stood waiting to save any who might rise from the water. There were six officers and seventy men on the *Eber* when she struck the reef, and of these five officers and sixty-six men were lost. This was about six o'clock in the morning.

During the excitement attending that calamity the other vessels had been for the time forgotten, but it was soon noticed that the positions of several of them had become more alarming. The *Adler* had been swept across the bay, close to the reef, and in half an hour she was lifted on top of the reef and turned completely over on her side. Nearly every man was thrown into the water, but as almost the entire hull was exposed, all but twenty succeeded in regaining her deck, and the remainder were rescued toward the close of the day when almost exhausted.

Just after the *Adler* struck, the attention of every one was directed toward the *Nipsic*. She was standing off the reef with her head to the wind, but the three anchors which she had out at the time were not holding; and orders were given to attach a hawser to a heavy eight-inch rifle on the forecastle and throw the gun overboard. As the men were in the act of doing this, the *Olga* bore down on the *Nipsic* and struck her amidships with awful force. Her bowsprit passed over the side of the *Nipsic*, and, after carrying away one boat and splintering the rail, came in contact with the smokestack, which was struck fairly in the center and fell to the deck with a crash like thunder. For a moment it was difficult to realize what had happened, and great confusion followed. The iron smokestack rolled from side to side with every movement of the vessel, until finally heavy blocks were placed under it. By that time the *Nipsic* had swung around and was approaching the reef, and it seemed certain that she would go down in the same way as had the *Eber*. Captain Mullan saw that any further attempt to save the vessel would be useless, so he gave the orders to beach her. She had a straight course of about two hundred yards to the sandy beach in front of the American consulate, where she stuck and stood firm.

Two attempts to lower boats were failures and every man crowded to the forecastle. A line was thrown, double hawsers were soon made fast from the vessel to the shore, and the natives and others gathered around the lines, where the voices of officers shouting to the men on deck were mingled with the loud cries and singing of the Samoans. One by one, and in a very orderly manner, the men of the *Nipsic* came down the hawsers toward the shore, but many would never have reached it, had it not been for the assistance of the Samoans, who, at the peril of their lives, stood in the boiling torrent, grasping those whose hold was broken from the rope.

Meanwhile, the four large men-of-war, *Trenton*, *Calliope*, *Vandalia*, and *Olga*, were still afloat and in a comparatively safe position; but about ten o'clock the *Trenton* was seen to be in a helpless condition; her rudder and propeller were both gone, and there was nothing but her anchors to hold her up against the unabated force of the storm. The *Vandalia* and *Calliope* were also in danger, drifting back toward the reef near the point where lay the wreck of the *Adler;* and they came closer together every minute, until finally the English ship struck the *Vandalia* and tore a great hole in her bow. Then Captain Kane of the *Calliope* determined to try to steam out of the harbor as his only hope, and he at once cut loose from all his anchors. The *Calliope's* head swung around to the wind and her engines were worked to their utmost power. Great waves broke over her bow and she gained headway at first only inch by inch, but her speed gradually increased until it became evident that she could leave the harbor. This manoeuvre of the British ship is regarded as one of the most daring in naval annals — the one desperate chance offered her commander to save his vessel and the three hundred lives aboard.

The *Trenton's* fires had gone out by that time, and she lay helpless almost in the path of the *Calliope*. The decks were swarming with men, but, facing death as they were, they recognized the heroic struggle of the British ship, and a great shout went up from aboard the *Trenton*. "Three cheers for the *Calliope !*" was the sound that reached the ears of the British tars as they passed out of the harbor in the teeth of the storm; and the heart of every Englishman went out to the brave American sailors who gave that parting tribute to the Queen's ship.

When the excitement on the *Vandalia* which followed the collision with the *Calliope* had subsided, it was determined to beach the vessel, and straining every means at hand to avoid the dreaded reef, she moved slowly across the harbor until her bow stuck in the sand, about two hundred yards off shore and probably eighty yards from the stern of the *Nipsic*. Her engines were stopped and the men in the engine-room and fire-room below were ordered on deck. The ship swung around broadside to the shore, and it was thought at first that her position was comparatively safe, as it was believed that the storm would abate in a few hours, and the two hun-

14

dred and forty men on board could be rescued then; but the wind seemed to increase in fury, and as the hull of the steamer sank lower the force of the waves grew more violent, yet no one on shore was able to render the least aid.

These terrible scenes had detracted attention from the other two men-of-war still afloat; but about four o'clock in the afternoon the positions of the *Trenton* and *Olga* became most alarming. The flagship had been in a helpless condition for hours, being without rudder or propeller, while volumes of water poured in through her hawser-pipes. Men never fought against adverse circumstances with more desperation than the officers and men of the *Trenton* displayed during those hours, yet the vessel was slowly forced over toward the eastern reef. Destruction seemed imminent, as the great vessel was pitching heavily, and her stern was but a few feet from the reef. This point was a quarter of a mile from shore, and if the *Trenton* had struck the reef there, it is probable that not a life would have been saved. A skilful maneuver, suggested by Lieutenant Brown, saved the ship from destruction. Every man was ordered into the port rigging, and the compact mass of bodies was used as a sail. The wind struck against the men in the rigging and forced the vessel out into the bay again. She soon commenced to drift back against the *Olga*, which was still standing off the reef and holding up against the storm more successfully than any other vessel in the harbor had done, and in spite of every effort on the part of both ships a collision took place which severely damaged both. Fortunately, the vessels drifted apart, whereupon the *Olga* steamed ahead toward the mud-flats in the eastern part of the bay, and was soon hard and fast on the bottom. Not a life was lost, and several weeks later the ship was hauled off and saved.

The *Trenton* was now about two hundred feet from the sunken *Vandalia*, and seemed sure to strike her and throw into the water the men still clinging to the rigging. It was now after five o'clock, and the daylight was beginning to fade away. In a half hour more, the *Trenton* had drifted to within a few yards of the *Vandalia's* bow, and feelings hard to describe came to the hundreds who watched the vessels from the shore.

Presently the last faint rays of daylight faded away, and night came down upon the awful scene. The storm was still raging with as much fury as at any time during the day. The poor creatures who had been clinging for hours to the rigging of the *Vandalia* were bruised and bleeding; but they held on with the desperation of men who were hanging between life and death. The ropes had cut the flesh on their arms and legs, and their eyes were blinded by the salt spray which swept over them. Weak and exhausted as they were, they would be unable to stand the terrible strain much longer. The final hour seemed to be upon them. The great, black hull of the *Trenton* was almost ready to crash into the stranded *Vandalia* and grind her to atoms. Suddenly a shout was borne across the waters. The sound of four hundred and fifty voices was heard above the roar of the tempest. "Three cheers for the *Vandalia!*" was the cry that warmed the hearts of the dying men in the rigging.

The shout died away upon the storm, and there arose from the quivering masts of the sunken ship a response so feeble it was scarcely heard upon the shore. Every heart was melted to pity. "God help them!" was passed from one man to another. The cheer had hardly ceased when the sound of music came across the water. The *Trenton's* band was playing "The Star-spangled Banner." The thousand men on sea and shore had never before heard strains of music at such a time as that. An indescribable feeling came over the Americans on the beach who listened to the notes of the national song mingled with the howling of the storm.

But the collision of the *Trenton* and *Vandalia*, instead of crushing the latter vessel to pieces, proved to be the salvation of the men in the rigging. When the *Trenton's* stern finally struck the side of the *Vandalia*, there was no shock, and she swung around broadside to the sunken ship. This enabled the men on the *Vandalia* to escape to the deck of the *Trenton*, and in a short time they were all taken off.

The storm had abated at midnight, and when day dawned there was no further cause for alarm. The men were removed from the *Trenton* and provided with quarters on shore.

During the next few days the evidences of the great disaster could be seen on every side. In the harbor were the wrecks of four men-of-war: the *Trenton*, *Vandalia*, *Adler*, and *Eber*; and two others, the *Nipsic* and *Olga*, were hard and fast on the beach and were hauled off with great difficulty. The wrecks of ten sailing-vessels also lay upon the reefs. On shore, houses and trees were blown down, and the beach was strewn with wreckage from one end of the town to the other.

Ever since men began to go to sea lights have been placed on shore to guide them to a landing-place: but in early times these were nothing more than fires on headlands, kindled, perhaps, by the wives and children of the captain and his crew of neighbors, when these mariners were expected home. These friendly services became a little more systematic when merchants began to risk their property on the water: and on the shores of the Mediterranean, which we have found to be the cradle of civilized navigation and trade, harbor-beacons were erected in very early times as guides to a safe anchorage.

The giant statue known as the Colossus, at Rhodes, is supposed to have been used as a beacon and lighthouse, a fire burning in the palm of its uplifted colossal hand at night. Although the account of the Colossus is only a matter of guesswork, it is historically true that in those ages of ignorant heedlessness of the need of beacons, a lighthouse was built so grand in proportions, so enduring in character, that it became known as one of the Seven Wonders of the World, and outlived all the others, save the Pyramids, by centuries, and in some ways has never been excelled by any similar structure in modern times, unless it be by our mammoth marble monument to Washington. This was the lighthouse built on the little island of Pharos by Ptolemy Philadelphus, king of Egypt, two hundred and eighty years before Christ, to guide vessels into the harbor of Alexandria. From all descriptions, it must have closely resembled our Washington monument; for it was built of white stone, was square at the base, and tapered toward the apex. Open windows were near its top, through which the fire within could be seen for thirty miles by vessels at sea.

The destruction of these beacons in the general smash and ruin that seem to have overtaken the world when the Roman empire went to pieces is only indicative of the way the darkness of barbarism returned and enveloped the minds as well as the works of men, until light broke through the clouds again with the rise of organized sea-powers in Western Europe. Then beacons were gradually rebuilt, but in almost all cases by private hands — the feudal lords of coast estates, the master or authorities of sea-ports, the monks in monasteries near dangerous landings, and now and then the king at his principal port, setting up marks for steering by day and lighting fires on dark nights. Most of the latter were hardly more than tar-barrels, which would burn brightly in a gale, and the better class were towers of stonework, on top of which a mass of coal was ignited in an iron cage, and kept stirred into brightness by a watcher.

It was an easy matter to imitate such beacons, and wreckers would often set up false lights. Many a fearful tradition has come down of the doings of wreckers, not only in England and Spain, but in America and in the East. One of their tricks, when they saw a ship approaching in the evening, was to hang a lantern upon a horse's neck, and let him graze, well-hobbled, along the beach. This would appear like the rocking of a lantern on a vessel at rest—what is called a riding or anchor light; and, deceived by this promise of a safe anchorage, the stranger would not discover that he had been cheated until his keel struck a reef or sandbar, and the pirates had begun their villainous attack. It is said to have been a device of this kind which caused the wreck in 1812, on the Carolina coast,—whose islands and lagoons are reputed to have been infested by such ruffians, there known as "bankers,"—of a vessel carrying the beautiful Theodosia Burr, daughter of Aaron Burr, and wife of Governor Alston of South Carolina. Her death at the hands of these men is illustrated on page 172.

During the reign of Henry VIII of England, an association of mariners called, in short, the Guild of the Trinity was chartered and given various powers and privileges in connection with the newly instituted royal navy and dockyards. It encouraged coast-lights, and in 1573 Queen Elizabeth formally placed authority to erect and govern lighthouses and coast beacons in the hands of this corporation, and there it remains to this day; for its headquarters, Trinity House, on Tower Hill, in London, are a recognized office of the British government, answering to our Lighthouse Board.

It was not long before it encouraged the founding of a permanent light on Eddystone Shoals, a group of reefs near Plymouth, exceedingly dangerous because they lie precisely in the track of ships bound up or down the English Channel, yet almost invisible. Upon the mere standing-room afforded by the crest of this rock, Sir William Winstanley managed to erect, two hundred years ago, a tower of wood and iron trestle-work, bolted to its foundation and carrying a glass room or lantern containing a coal-grate, eighty feet above low-water mark. This was completed in 1698. One winter's experience convinced him that it needed strengthening, and in 1699 a case of masonry was built about the tower, and made solid to the height of twenty feet, while the whole structure was increased to the height of one hundred and twenty feet. Then, it is related, Winstanley boasted that the sea had not strength enough to tear it down, and all England rejoiced in so noble a beacon; but we now know that the construction was faulty, in its large diameter, polygonal outline, excess of ornament, and lack of weight. While Sir William was within it making repairs, four years later, the memorable hurricane of November 20, 1703, swept the coast, and left scarcely a trace of the tower. Its value had been proved, however, and it was replaced, in 1706, by a straight-sided tower of oaken timbers, weighted in their lower courses by stone. This was designed by an engineer named Rudyerd, and lasted until burned down in 1755; and engineers say it was better for its place than was the round, solid-based stone tower of Smeaton that followed it, and became so celebrated. This was finished in three years, and in 1760 was lighted, not by a fire, as of old, but by candles—the first use of such an illuminant. This truly illustrious

THE WRECK OF THE FIRST MINOT'S LEDGE LIGHTHOUSE.

lighthouse remained until a few years ago, when it became so racked by the assaults of the sea as to be unsafe. It was then replaced by the one that stands there to-day, rivaling its magnificent neighbor on the Biscay shore opposite, the lighthouse of Carduan, which was built to support a bonfire of oak, but has remained to be lighted successively by oil-lamps, by gas-burners, and finally by electricity.

A somewhat similar history belongs to some of the lighthouses on this side of the Atlantic. The first one regularly set up in the United States was that on the north side of the entrance to Boston harbor, erected in 1716; but many others go back to Colonial days—that on Sandy Hook, for instance. Perhaps the most interesting history is attached to the light on Minot's Ledge, in Boston harbor. This is a dangerous reef, concealed at high water and so exposed that the problem of lighting it was much the same as that presented at Eddystone, Bell Rock, Dhu Heartach, and other well known islets on the British coast.

The first lighthouse on Minot's Ledge was built in 1848, and was an octagonal tower resting on the tops of eight wrought-iron piles sixty feet high, eight inches in diameter, and sunk five feet into the rock.

A SCREW-PILE OCEAN LIGHTHOUSE.

These piles were braced together in many ways, and, as they offered less surface to the waves than a solid structure, the lighthouse was considered by all authorities upon the subject to be exceptionally strong. Its great test came in April, 1851. On the fourteenth of that month, two keepers being in the lighthouse, an easterly gale set in, steadily increasing in force. . . . On Wednesday, the sixteenth, the gale had become a hurricane; and when at times the tower could be seen through the mists and sea drift, it seemed to bend to the shock of the waves. At four o'clock that afternoon an ominous proof of the fury of the waves on Minot's Ledge reached the shore — a platform which had been built between the piles only seven feet below the floor of the keeper's room. The raging seas, then, were leaping fifty feet in the air. Would they reach ten feet higher? — for if so, the house and the keepers were doomed. Nevertheless, when darkness set in the light shone out as brilliantly as ever, but the gale seemed, if possible, then to increase. What agony those two men must have suffered! How that dreadful abode must have swayed in the irresistible hurricane, and trembled at each crashing sea! The poor unfortunates must have known that if those seas, leaping always higher and higher, ever reached their house, it would be flung down into the ocean, and they would be buried with it beneath the waves.

To those hopeless, terrified watchers the entombing sea came at last. At one o'clock in the morning the lighthouse bell was heard by those on shore to give a mournful clang, and the light was extinguished. It was the funeral knell of two patient heroes.

Next day there remained on the rock only eight jagged iron stumps.

Thus, everywhere, and in all latitudes, the beacons and wooden towers and huge pyramids of long ago have given place to slender spires of solid masonry, holding powerful signals perhaps hundreds of feet above the waves, and visible as far as the curve of the earth's surface will permit. Yet in place of the sturdy bonfire of oak, or the huge iron cage full of coals, there is only a single lamp, whose rays are gathered by deep reflectors into a compact bundle of unwasted rays, and doubled and redoubled by rows of magnifying lenses until they can dart to the furthest horizon in a strong beam of steady light. No longer does the mariner trust to his wife to kindle the tar-barrel to guide him home. He knows that nowhere is his government more watchful of its subjects than in its lighthouse service, and that he may trust to having that bright signal to welcome him in the darkness, as well as he can trust his own eyes to see it. The United States alone expends over $2,500,000 annually in looking after her lighthouses, lightships, and buoys.

Indeed, these beacons have become so thickly planted that it has been found necessary to distinguish between them in order to avoid mistaking one for another. At first this was done by doubling, as in the case of New York's "Highland Lights," or the twin lights of Thatcher's Island off Cape Ann, or even trebling them as at Nauset, on Cape Cod, but now the display is made to vary. Thus some of them are simply fixed white lights; some are white and revolve — the whole lantern on the summit of the tower being turned on wheels by machinery, and the flame disappears for a longer or shorter time; while others are white "flash" lights, glancing only for an instant, and then lost for a few seconds, or giving a long wink and then a short one with a space of darkness between. Some lighthouses show a steady red light; others, alternate red and white. By these colors and varying periods of appearance and disappearance (noted on charts, and published by the government in a general seaman's guide called the "Coast Pilot"), navigators know which light they

THE LIGHTHOUSE AT ST. AUGUSTINE, FLORIDA.

are looking at when several are in sight. For daylight recognition the towers may be painted half black and half white, or in stripes or bands or spirals, like the big barber's pole in front of St. Augustine, Florida.

It is impossible here to describe in detail the beautiful machinery by which the rays from the large but simple argand kerosene lamp are condensed into a single beam and projected through the Fresnel system of condensers and lenses, and by which the revolution and "flashing" are effected. Petroleum has superseded all other oils for general use, but electricity is now being extensively employed in the illumination of coast lights, especially in France, where they are introducing new principles, such as producing lightning-like flashes with a certain recognized regularity, and waving stupendous search-light beams in the sky, so that the approach to the coast may be seen when the land and lighthouse themselves are still below the horizon. If you have an opportunity to go into the lantern of a lighthouse, by all means take advantage of it; and if you can be there when a storm is raging, or when, on some misty night, the lantern is besieged by migrating birds, you will never forget the scene.

On some especially dangerous — because hidden — shoals, reefs, or bars, like those off Nantucket or the extreme point of Sandy Hook, it may be out of the question or bad policy to erect a lighthouse. Here its place is taken by anchoring a stout vessel, built to withstand the severest weather, and arranged to carry lanterns at its mastheads.

These are called "lightships," and they are manned by a crew of keepers who have a very monotonous and uncomfortable time of it; yet in some cases men have spent twenty years or more in the service.

The most desolate and dangerous lightship station is that of No. 1, Nantucket. "Upon this tossing island, out of sight of land, exposed to the fury of every tempest, and without a message from home during all the stormy months of winter, and sometimes even longer, ten men, braving the perils of wind and wave, and the worse terrors of isolation, trim the lamps whose light warns thousands of vessels from certain destruction, and hold themselves ready to save life when the warning is vain."

Seven years ago Mr. Gustav Kobbé, and the artist, William Taber, spent several days on the lightship and gave a graphic account of the life there, which I wish I were able to quote in full.

The anchorage is twenty-four miles out at sea beyond Sankaty head, at the extremity of the shoals and rips which make all that space of water beyond the visible coast of Nantucket fatal to ships, hundreds of which are known to have been beaten to pieces on its treacherous bars. She is moored to a 6500-pound mushroom anchor by a chain two inches in thickness, yet she has been torn adrift twenty-three times, and has wandered widely before returning or being overtaken.

"No. 1, Nantucket New South Shoals," to quote Mr. Kobbé,

is a schooner of two hundred and seventy-five tons, one hundred and three feet long, with twenty-four feet breadth of beam, and stanchly built of white and live oak. She has two hulls, the space between them being filled through holes at short intervals in the inner side of the bulwarks with salt. . . . She has fore-and-aft lantern-masts seventy-one feet high, including topmasts, and directly behind each of the lantern masts a mast for sails forty-two feet high. Forty-four feet up the lantern-masts are day-marks, reddish brown hoop-iron gratings, which enable other vessels to sight the lightship more readily. The lanterns are octagons of glass in copper frames five feet in diameter, four feet nine inches high, with the masts as centers. Each pane of glass is two feet long and two feet three inches high. There are eight lamps, burning

LIGHTSHIP NO. 1, NANTUCKET NEW SOUTH SHOALS.

a fixed white light, with parabolic reflectors in each lantern, which weighs, all told, about a ton. Some nine hundred gallons of oil are taken aboard for service during the year. The lanterns are lowered into houses built around the masts. The house around the main lantern-mast stands directly on the deck, while the foremast lantern-house is a heavily-timbered frame three feet high. This is to prevent its being washed away by the waves the vessel ships when she plunges into the wintry seas. When the lamps have been lighted and the roofs of the lantern-houses opened,—they work on hinges, and are raised by tackle,—the lanterns are hoisted by means of winches to a point about twenty-five feet from the deck. Were they to be hoisted higher they would make the ship top-heavy.

A conspicuous object forward is the large fog-bell swung ten feet above the deck. The prevalence of fog makes life on the South Shoal Lightship especially dreary. During one season fifty-five days out of seventy were thick, and for twelve consecutive days and nights the bell was kept tolling at two-minute intervals.

The actual work to be done is small, the daily cleaning of the lamps requiring only two or three hours, and other chores being very light, and the

men nearly die of loneliness and "nothing to do." It is pathetic to read
how intense and friendly an interest they take in a single red buoy anchored
near them; and they admit that fog is dreaded more because it hides this
neighbor than for any other reason.

Mr. Kobbé tells us that the emotional stress under which this crew labors
can hardly be realized by any one who has not been through a similar
experience.

The sailor on an ordinary ship has at least the inspiration of knowing that he is bound for
somewhere; that in due time his vessel will be laid on her homeward course; that storm and fog
are but incidents of the voyage; he is on a ship that leaps forward full of life and energy with every
lash of the tempest. But no matter how the lightship may plunge and roll, no matter how strong
the favoring gales may be, she is still anchored two miles southeast of the New South Shoal.

CLEANING THE LAMPS ON A LIGHTSHIP.

Besides enduring the hardships incidental to their duties aboard the lightship, the South
Shoal crew have done noble work in saving life. While the care of the lightship is considered
of such importance to shipping that the crew are instructed not to expose themselves to dangers
outside their special line of duty, and they would therefore have the fullest excuse for not risking
their lives in rescuing others, they have never hesitated to do so. When, a few winters ago,
the *City of Newcastle* went ashore on one of the shoals near the lightship, and strained herself so
badly that although she floated off, she soon filled and went down stern foremost, all hands,
twenty-seven in number, were saved by the South Shoal crew and kept aboard of her over two
weeks, until the story of the wreck was signaled to some passing vessel and the lighthouse tender
took them off. This is the largest number saved at one time by the South Shoal, but the
lightship crew have faced great danger on several other occasions.

This is, perhaps, the extreme picture of lightship life, but apart from
the prolonged isolation and continuous roughness of the water, the experi-

ences of the men off Sandy Hook and elsewhere
are not greatly removed from it, and no philanthropy
is more worthy of support than that which seeks to
mitigate the loneliness of these exiles by providing
them with reading matter. The Lighthouse Board
provides a small circulating library for these ships,
and contributions of books and files of illustrated
periodicals will be gratefully received and put to
good use by the Superintendent of the Lighthouse
Service in Washington.

But there are times — and they occur very frequently in northern waters — when fogs which no
light can penetrate envelop sea and coast, and that
is the most dangerous of all times to an approaching
ship. The only means by which a warning can
be given, in such an emergency, is by sound. In
many places bells are rung, but often the place to
be avoided is so situated that the roar of the surf

THE FOG-BELL.

would drown a bell's note, and then fog-horns are blown. These fog-horns
are of a size so immense, and voices so stentorian, that it requires a steam
engine to blow them, and they utter a booming, hollow blast, a dismal note
as we hear it when we are safe on the land, but sweet to the anxious captain
whose vessel is laboring through the gloom under close-reefed topsails, and
uncertain of her exact position. One kind of these horns is very complicated
in its structure, and screeches in a rough, broken blare, a note far-reaching
beyond any smooth, whistling sound that could be made. This shriek is so
hideous, so ear-splitting, when heard near at hand, that no name bad enough
to express it could be found; so its inventors went to the other extreme, and
called it a siren, after those most enchanting of sweet singers who tried to
entice Ulysses out of his course. This name is opposite in a double sense,
indeed, for the sirens of old lured sailors to wreck, while our siren hoarsely
bids them keep off. Finally, buoys, which at first were simply tight casks,
but now are usually made of boiler-iron, are anchored on small reefs, to
which are hung bells, rung constantly by the tossing of their support; and
on other reefs buoys are fixed having a hollow cap so arranged that when a
big wave rushes over, it shuts in a body of air, under great and sudden
pressure, which can only escape through a whistle in the top of the cap, ut-
tering a long warning wail to tell its position.

It is in such times as this that the pilot comes out strong.

A pilot is a man who has made himself thoroughly acquainted with

A SIREN RIGGED UPON A MERCHANT STEAM-SHIP.

certain waters where navigation is dangerous, and who is licensed by some
proper authority, after training and examination, to direct vessels in safety
in entering harbors or passing through other intricate places. A ship-cap-
tain may be an excellent navigator, but he is not expected to know every
rock and sandbar crouching under the waves, and all the twistings and
turnings of the entrance and channel of a foreign harbor, especially as these
channels are subject to constant change. In this country, indeed, although
coasting-vessels may refuse a pilot, the law will not permit captains coming
from or bound to a foreign port to do so; and if any accident happens
when no pilot is aboard the insurance money will not be paid, and the ship's
officers may be punished.

Pilots, then, are important men and are able to charge very high prices
for their services (generally rated according to the draft of the vessel), and
their profession is so organized and guarded that not only must a man be
thoroughly competent, but he must wait for a vacancy in the regular
number before he will be admitted to their ranks.

Their method of work is very exciting. A dozen or so together will
form the crew of a trim, stanch schooner, provisioned for a fortnight or
more, which can outsail anything but a racing yacht, and is built to ride
safely through the highest seas. A few steamers are coming into use, but
the procedure is much the same. You will now and then see one of these
beautiful little vessels sailing up the quiet harbor, threading its way
through the black steamers and sputtering tug-boats and great ships, as a
shy and graceful girl walks among the guests at a lawn party, and you
know from its air as well as the big number on its white mainsail that it is a
pilot boat, even if it does not carry the regular pilot-flag, which in the
United States is simply the "union" or starry canton of the ensign.

But these fine schooners and the brave men they carry are rarely in
port. Their time is spent far in the offing of the harbor, cruising back and

forth in wait for incoming ships, and the New York pilots often go two and three hundred miles out to sea, and in storms may be blown much farther away. Other pilot-boats are waiting also, and the lookout at the reeling

BURNING A "FLARE" ON A PILOT-BOAT.

mast-head must keep the very keenest watch upon the horizon. Suddenly he catches sight of a white speck which his practised eye tells him is a ship's top-sails, or of a blur upon the sky that advertises a steamer's approach. The schooner's head is instantly turned toward it, and all the canvas is crowded on that she will bear, for away off at the right a second pilot-boat, well down, is also seen to be aiming at the same point and trying hard to win.

The first pilots of New York harbor were stationed at Sandy Hook, and visited incoming vessels in whale-boats; and many a stately British frigate or colonial trader was forced to wait anxiously outside the bar, rolling and tossing in the sea-way, or tacking hither and yon, hoping for a glimpse of that tiny speck where flashing oars told of the coming pilot. It is in this way, as the late Mr. J. O. Davidson, the artist, who knew all about such things, told us in "St. Nicholas" (January, 1890), that many vessels are still met, off some of our smaller harbors, and at the mouth of the Mississippi River. There the waters of the great river pouring into the Gulf of Mexico through the Port Eads Jetties make a turbulent swell with foam-crested billows that roll the stoutest ship's gunwale under, even in calm weather; yet the little whale-boats, swift and buoyant, dash out bravely in a race for the sail on the distant horizon, for there are two pilot-stations at the Jetties, and it is "first come first engaged."

Sometimes, on the other hand, it is the ship that looks for the pilot, cruising about with the code-letters P T flying from her signal-halyards in token of her need. She may even run past a pilot-boat in the night and get into danger without being aware of it. To prevent this, says Mr. Davidson, the pilots burn what is known as a "flare" or torch, consisting of a bunch of cotton or lamp-wick dipped in turpentine, on the end of a short handle. It burns with a brilliant flame, lighting up the sea for a great distance and throwing the sails and number of the pilot-boat into strong relief against the darkness. On a dark clear night, the reddish glare which the signal projects on the clouds looks like distant heat lightning.

Having sighted his vessel, the pilot whose turn it is to go on duty hurries below and packs the valise which contains such things as he wishes to take home, for this is his method of going ashore; and when he has departed, if he is the last one of the pilot-crew, the little vessel returns herself to port in charge of the sailing-master, cook, and "boy," to refit and take on a new set of men.

The storm may be howling in the full force of the winter's fury, and the waves running "mountain-high," but the pilot must get aboard by some means. It is rough weather indeed when his mates cannot launch their yawl and row him to where he can climb up the stranger's side with the aid of a friendly rope's end.

Yet frequently this is out of the question. Then a "whip" is rigged beyond the end of a lee yard-arm, carrying a rope rove through a snatch-block, and having a noose at its end. The steamer slows her engines, or

A PILOT BOARDING A STEAMER.

the ship heaves to, and the pilot-schooner, under perfect control, runs up under the lee of the big ship, as near as she dares in the gale. Then, just at the right instant, a man on the ship's yard hurls the rope, it is caught by the schooner, the pilot slips one leg through the bowline-noose, and a second afterward the schooner has swept on and he is being hoisted up to the yard-arm, but generally not in time to save himself a good ducking in the coaming of some big roller. Going on shipboard in this fashion is not favorable to an imposing effect; nevertheless, the pilot is welcomed by both crew and passengers, who admire his courage and trust his skill, but smile at the high hat beloved of all pilots.

Now the pilot is master — stands ahead of the captain even — and his orders are absolute law. He inspects the vessel to form his opinion of how she will behave, and then goes to the wheel or stands where best he can give his orders to the steersman and to the men in the fore-chains heaving the sounding lead. He must never abandon his post, he must never lose his control of the ship, or make a mistake as to its position in respect to the lee-shore, or fail to be equal to every emergency. If it is too dark and foggy and stormy to see, he must feel; and if he cannot do this he must have the faculty of going right by intuition. To fail is to lose his reputation if not his life. This is what is expected of pilots, and this is what they actually do in a hundred cases, the full details of any one of which would make a long and thrilling tale of adventurous fighting for life.

It is to help pilots and navigators of all sorts to avoid the perils that beset them that governments not only spend large sums in surveying coasts and harbors, publishing charts and descriptions, and maintaining lighthouses and lightships, but mark out bars and channels with floating guides, and their borders with shore-beacons and "ranges," to form so many finger-posts for the right road. Were it not for these sign-posts no ship could safely enter any commercial harbor in the world; and it will be valuable to quote somewhat from an article, with capital illustrations, written for "St. Nicholas" (March, 1896) by an officer of the United States Navy, Lieut. John M. Ellicott, since it describes how the long, winding approaches to one of the greatest ports of the world are marked out by day and by night — I mean the harbor of New York.

Suppose, then, that we are on a big transatlantic steamer approaching the United States from Europe. . . . Having secured his pilot, it is the captain's next aim to make a "land-fall" —that is to say, he wishes to come in sight of some well-known object on shore, which, being marked down on his chart, will show him just where he is and how he must steer to find the entrance to the harbor.

A special lighthouse is usually the object sought, and in approaching New York harbor it is customary for steamers from Europe to first find, or sight, Fire Island Lighthouse. This is on a sandy island near the coast of Long Island. When, therefore, the liner steams in sight of Fire Island Light she hoists two signals, one of which tells her name and the other the welfare of

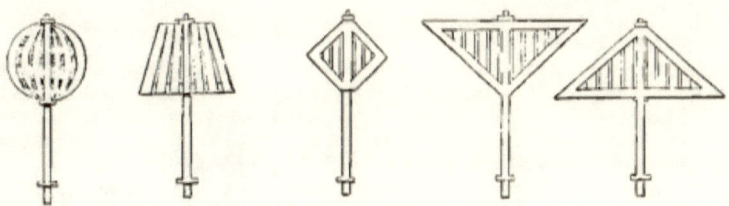

DAY-MARKS IN NEW YORK HARBOR.

those on board. The operator then telegraphs to the ship's agents in New York that she has been sighted, and that all on board are well or otherwise. [Other despatches go to the newspapers, who have observing stations and telegraph arrangements here and at Sandy Hook.]

The ship's course is then laid to reach the most prominent object at the harbor entrance, in this case Sandy Hook Lightship. She is easily recognized. The course from this lightship to the harbor entrance is laid down on the chart "west-northwest, one quarter west," and, steering this course, a group of three buoys is reached. One is a large "nun," or cone-shaped, buoy, painted black and white in vertical stripes; another has a triangular framework built on it, and in the top of this framework is a bell which tolls mournfully as the buoy is rocked; while the third is surmounted by a big whistle. . . . These mark the point where ocean ends and harbor begins, and can be found in fair weather or in fog by their color and shape, or noise. They are the mid-channel buoys at the entrance to Gedney Channel, the deep-water entrance to New York harbor. Here it may be noted that mid-channel buoys in all harbors in the United States are painted black and white in vertical stripes, and, being in mid-channel, should be passed close to by all deep-draft vessels. At this point the pilot takes charge.

Ahead the water seems now to be dotted in the most indiscriminate manner with buoys and beacons, and on the shores around the harbor, far and near, there seem to be almost a dozen lighthouses. If, however, you watch the buoys as the pilot steers the ship between them, you will soon see that all those passed on the right-hand side are *red*, and all on the left are *black*. Where more than one channel runs through the same harbor, the different channels are marked by buoys of different shapes. Principal channels are marked by "nun" buoys, secondary channels by "can" buoys, and minor channels by "spar" buoys.

Gedney Channel is a short, dredged lane leading over the outer bar, or barrier of sand, which lies between harbor and ocean. Its buoys are lighted at night by electricity, through submarine cables, the red ones with red lights, the black ones with white lights. Moreover, a little lighthouse off to the left, known as Sandy Hook Beacon, has in its lamp a red sector which throws a

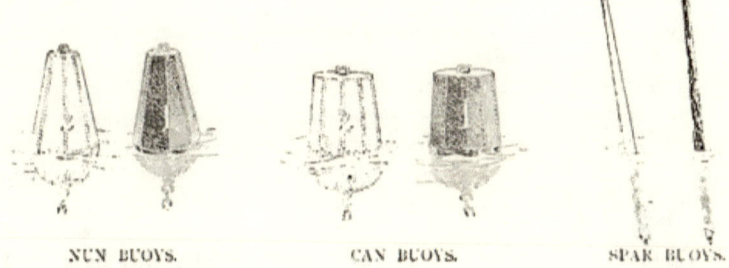

NUN BUOYS.   CAN BUOYS.   SPAR BUOYS.

red beam just covering Gedney Channel. Thus this channel can be passed through in safety by night as well as by day. If it is night the pilot knows when he is through it by the change of color in Sandy Hook Beacon light from red to white. Then he looks away past that light to his left for two fixed white lights on the New Jersey shore and hillside, known as Point Comfort Beacon and Waackaack Beacon, for he knows that by keeping them in range, that is to say, in line with one another and himself, and by steering toward them he is in the main ship channel. By day the main ship channel buoys would guide him, as in Gedney Channel, but at night these buoys are not lighted.

Only a short distance is now traversed when the ship comes to a point where two unseen channels meet. This is indicated by a buoy having a tall spindle, or "perch," surmounted by a latticed square. From here, if she continues on her course, she will remain in the main ship-channel, which, although deeper, is a more circuitous route into port; so, if she does not draw too much water, she is turned somewhat to the right, and, leaving the buoy with the perch and square on her right, because it is red, she is steered between the buoys which mark Swash Channel. If it were night this channel would be revealed by two range-lights on the Staten Island shore and hillside, known as Elm Tree Beacon and New Dorp Beacon, both being steady-burning, white lights; but we are entering by daylight, and when half-way through Swash Channel we notice a buoy painted red and black in horizontal stripes. To this is given a wide berth by the pilot. It is an "obstruction" buoy marking a shoal spot or a wreck.

OBSTRUCTION BUOY.

Channel buoys are all numbered in sequence from the sea inward, the red ones with even, and the black ones with odd numbers, and the larger ones are anchored with "mushrooms" while the smaller have "sinkers" of iron or stone. They are made of iron plates in water-tight compartments, so that if punctured by an over-running ship or some other accident, they will not be likely to sink. In harbors where ice forms in winter, large summer buoys are replaced in winter by a smaller sort less liable to be torn adrift. Buoys do go adrift, however, now and then, and sometimes take a voyage across the ocean or far down the coast before they can be found by the tenders of the Lighthouse Service, which is constantly looking after these and other marks. Lieutenant Ellicott tells us that all changes in the position of buoys or lightships, or the placing of new buoys to mark a change of channel, or an obstruction, are published promptly in pamphlets called "Notices to Mariners," which are distributed as quickly as possible through well organized means of communication. A few years ago one of the largest of our handsome new cruisers was approaching New York harbor from the West Indies in a light fog. Sandy Hook Lightship had been found, the usual course laid for Gedney Channel, and the ship was steaming onward at full speed, her captain, having been inspector of that very lighthouse district but a short time before, feeling that he knew his way into that port better than the most experienced pilot. Presently, however, he was startled by the alarming cry of *breakers ahead!* A large hotel also loomed up, and, as the ship was backed full speed astern, all hands realized that they had barely escaped running high and dry on Rockaway Beach. When the vessel got into port it was learned that Sandy Hook Lightship had been moved considerably from its old position, and that the notice of this change had failed to reach the captain of the cruiser before he sailed from the West Indies.

Shipwrecks still occur, however, in spite of lighthouses and sirens and buoys and coast-surveys; therefore we add to our precautions arrangements to help those cast away. Societies to save wrecked persons have existed, it is said, for many centuries in China, but in Europe they are hardly a hundred years old. The early humane societies, like that of Great Britain,

A WHISTLING BUOY, OUT OF COMMISSION.

placed life-boats and rescuing gear in certain shore towns, and organized crews, who promised to go out to the aid of any lost ship, and to take good care of the persons rescued.

In America, however, our coasts are so extensive, and so much of the dangerous part of them is far away from villages, or even a farmhouse, that the government has been obliged to do whatever was necessary. Thus came about the Life Saving Service, which now has its stations close together along our whole sea-coast, and upon the great lakes, covering more than ten thousand miles in all.

Each of these stations is a snug house on the beach, tenanted by a keeper and six men, all of whom are chosen for their skill in swimming, and in handling a boat in the surf—something every man who "follows the sea" cannot do successfully. Beaching a boat through surf is an art.

During all the season, from October till May, two men from each station are incessantly patrolling the beach at night, each walking until he meets the patrolman from the next station. No matter how foul the weather, these watchmen are out until daylight looking for disasters. The moment they discover a vessel ashore, or likely to become disabled, they summon their companions and hasten to launch their boat. These boats are of two kinds. On the lakes and on the steep Pacific coast is used the very heavy English life-boat, fitted with masts and sails if necessary, which a steam tug is required to tow to the scene of the wreck, unless it is close in shore. But upon our flat, sandy Atlantic beaches only a lighter kind of surf-boat, made of cedar, can be handled. This is built with air-cases at each end and under the thwarts, so that it cannot sink. The station men drag it on its low wagon to the scene of its use, unless horses are to be had, and when it is launched they sit at the six oars, each with his cork belt buckled around him, and his eye fixed on the steersman, who stands in the stern, ready to obey his slightest motion of command, for rowing through the angry waves that dash themselves on a storm-beaten beach is a matter requiring extraordinary skill and strength. Then, when the vessel is reached, comes another struggle to avoid being

PATROLMEN EXCHANGING THEIR CHECKS.

struck and crushed by the plunging ship, or the broken spars and rigging pounding about the hull. But skill and caution generally enable the crew to rescue the unfortunate castaways one by one, though frequently several trips must be made, in each one of which every surfman risks his life, and in many a sad case loses it; yet there is no lack of men for the service.

It is a common occurrence, however, that the sea will run so high that no boat could possibly be launched. Then the only possibility of rescue for the crew is by means of a line which shall bridge the space between the ship and the land before the hull falls to pieces. We read in old tales of wrecks of how some brave seaman would tie a light line around his waist, and dare the dreadful waves, and the more dreadful undertow, to save his comrades. If he got safely upon the beach, he drew a hawser on shore and made it fast. Now we do not ask this; but with a small cannon made for the pur-

SAVING A SAILOR BY MEANS OF THE BREECHES BUOY.

pose, a strong cord attached to a cannon-ball is fired over the ship, even though it be several hundred yards distant. Seizing this line as it falls across their vessel, the imperiled sailors haul to themselves a larger line, called a "whip," which they fasten in a tackle-block in such a way that a still heavier cable can be stretched between the wreck and the land and made fast.

Then by means of a small side-line and pulleys a double canvas bag, shaped like a pair of knee-breeches, is sent back and forth between the ship and the shore, bringing a man each time, until all are saved.' Should there be many persons on board, though, and great haste necessary, instead of the breeches-buoy a small covered metallic boat, called the life-car. is sent out, into which several persons may get at once. These varied means are so skilfully employed, that now hardly one in two hundred is lost of those whose lives are endangered on the American coasts.

THE SELF-RIGHTING LIFE-BOAT.

CHAPTER XI

FISHING AND OTHER MARINE INDUSTRIES

HE grandest sea-chase is that after the whale — the most gigantic of mammals, the most extraordinary in appearance and habits, and the most valuable to man, for the capture of one may mean ten times as much reward as the ivory of an elephant or the rarest otter-skin would afford, and perhaps a hundred times as much, if ambergris be found within its body.

Men have had the hardihood to chase these huge and often savage creatures in their own turbulent element, and with the most primitive weapons, ever since the art of navigation was acquired.

The Japanese and other Asiatics of the western shore of the North Pacific have dared to go out in rowboats and attack the largest whales since the origin of their traditions, and they had a method of entangling these leviathans in nets, which must have produced exciting scenes, as the monster struggled amid the bloody turmoil of waters to free himself from the innumerable connected cords that embarrassed his movements, rather than subdued his strength, until his life ebbed away through a hundred wounds.

On the Alaskan coast, and southward as far as Oregon, the Indians, and especially those of the Queen Charlotte Islands and the coasts of the Strait of Juan de Fuca, were accustomed, hundreds, perhaps thousands of years ago, to go far away into the ocean in their dug-out canoes, searching for and spearing the whales with lances made of flint or bone, having detachable barbed heads. These were attached to shafts by rawhide lines, and to the shafts were attached buoys of large inflated bladders. When the animal was struck, the heavy pole would drive the lancehead through the skin and then fall off. The barbs would not only hold the instrument there, but cause it to work deeper and deeper, and the whale, darting away or diving, would be so impeded by dragging the poles and buoys after him, that he would soon return to receive other darts, and so, between loss of blood and exhaustion, would ultimately be killed. It is extremely interesting to read the stories, gathered by early travelers from the lips of the Indians,—old Haidas or Makahs are living yet who have taken part in

AN OLD WHALER.

such nerve-testing canoe-chases,— of their fights with this gigantic foe far from land, and their hair's-breadth escapes; and it is not strange that many quaint ceremonies were devised to placate the waters and the power of the whale-god in advance, and to honor the sea-hunters when they returned.

The Greenlanders and Eastern Eskimos do not seem to have been able in their small skin boats to conquer the largest sort of whales, but the smaller ones, such as the white whale, fell to their spears in a similar way; and they took great pains to secure any dead or stranded cetacean that came within their reach, the bones of which were as valuable to them, in the absence of wood, as were the flesh, oil, and sinews.

The history of European whaling begins with the excursions of the Basques, who, as long ago at least as the tenth century, were accustomed to go out from their shore-towns in search of the southern right whale which frequents the Bay of Biscay and its offing. Doubtless their boats were small, half-decked, lugger-rigged "shyppes," carrying ten to fifteen men, and looking much like many of the Channel fishermen of to-day. This "fishery" supplied all Europe during the Middle Ages with the whalebone and oil which were among the luxuries of the rich at that time; but by the time the sixteenth century had arrived, whales had become so scarce in the Eastern Atlantic — where now they are almost extinct — that this industry must have ceased had not the Cabots shown the way to Newfoundland, to whose shores the Basques at once extended their voyages with excellent results, for in those days whales were commonly seen all along the American shore of the North Atlantic. But this remote fishery would have been too precarious and costly to be of great consequence had it not been for the early efforts, related in Chapter V, to find a passage to the East north of the continents. The earliest of these failed, but they brought back reports that the edge of the frozen sea abounded in whales, and men rushed into this newly discovered field of wealth, as, centuries later, they abandoned everything in headlong haste to go to the gold-fields of California, Australia, South Africa or the Yukon Valley.

The English did their best to monopolize the whale fishery at once, but the Dutch sent war-vessels, and in a fleet action almost at the edge of the ice in 1618 the Dutch conquered and opened the seas to all comers, while separate districts on the coast of Spitzbergen were assigned to each nationality. The English interest in the fishery declined, but the Dutch increased their attention to it, taking over one thousand whales each year. "About 1680," we read, "they had two hundred and sixty vessels and fourteen thousand seamen employed. Their fishery continued to flourish on almost as extensive a scale until 1770, when it began to decline, and finally, owing to the war, came to an end before the end of the century." The Germans were always associated with them, and continued to send a whaling fleet to

Barentz Sea and the Jan Mayen waters until 1873. Meanwhile the Greenland whaling-grounds had begun to attract British whalemen, followed by
the Danes in the early part of the last century; then this local industry fell

WHALERS TRYING OIL OUT OF BLUBBER.

off, but was revived about 1800, remained prosperous for many years, and is
still the support of Peterhead and a few other Scotch ports.

The abundance of whales near the coast was one of the prime inducements held out to colonists by North America, where whales often appeared close to the shore, or in harbors, as occasionally they do yet. Here, at first, whale-fishing was pursued wholly in rowboats launched from the beach. Many shore towns owned whaleboats and gear, each with its trained crew, and some kept a regular lookout, day by day, whose duty it was promptly to announce the appearance of any whale in the offing. Such was the case at Southampton, Long Island, for many years, and even now, occasionally, the town-crew there rushes away through the breakers after some stray visitor amid the excitement of the whole neighborhood, but this happens only at intervals of several years.

Before the end of the seventeenth century, however, the people of Nantucket Island were wont to cruise about the neighboring ocean for right whales, their voyage lasting six weeks or so as a rule, and now and then they would pick up a sperm whale. By the middle of the eighteenth century, however, sperm whaling was no longer profitable in the Northern Atlantic, while the Greenland grounds were overrun by European ships. American fishermen therefore turned their attention to the West, and for many years confined themselves mainly to catching the sperm whale, finding at first their best "grounds" in the south-middle Pacific. When the War of Independence came on, Nantucket was the leading whaling-port of the country, but all the New England towns were more or less engaged, and no less than three hundred and sixty vessels, large and small, were out. The Revolutionary War nearly destroyed the industry, and before it could well revive, the War of 1812 again subjected the whaling-ships to capture by English privateers and men-of-war all over the world. After that, however, they spread all over the Southern seas, and between 1840 and 1850 more than seven hundred were flying the flag of the United States.

The whaling vessels were large, stanch craft, usually bark rigged, distinguished by their old-fashioned shape, weather-stained, smoky appearance, enormous boats swinging from end to end of the ship from lofty davits, and try-works forward. They kept longer than any one else many relics of rigging, custom, and language, belonging to the seamanship of earlier generations; and no sea-peril could daunt either the vessel or its crew. They would sail on voyages lasting two or three years, and sometimes would circumnavigate the globe and return without having touched at a port. As a rule, however, they would gain part of a cargo, and then go to some port, ship it to London or New York, and refit for a new voyage. The profits of a trip were thus very great sometimes, but other trips were attended only by expense and misfortune.

The capture of whales in those days had more danger if not more ex-

A RACE FOR A WHALE.

citement than now, for the only method was by rowing after them, helped by the sails, in the 28-foot, double-ended rowboats made for the purpose (of which every vessel carried six or eight), and sinking into their vitals darts and lances until they died. They were then towed to the vessel's side, held by tackle from the yard-arms in a suitable position, and cut up. The oil in early days was packed in casks, but later has been run into iron tanks built into the hold, after having been tried out of the blubber in the great caldrons set in brick on the forward deck, which gave a whaler so peculiar an appearance, at all times, and would lead any one to suppose her on fire while the process of trying-out was going on, and the great volumes of black smoke caused by the use of whale-fat and waste as fuel were drifting to leeward.

One of the best accounts of a chase published is that by the late Temple Brown, of the United States Fish Commission, in an article in "The Century" for February, 1893, from which I am permitted to make an extract:

While cruising on the coast of New Zealand, one day about 11.30 A. M., the lookout at the main hailed the deck with: "Thar sh' b-l-o-w-s! Thar sh' b-l-o-w-s! Blows! B-l-o-w-s!"

"Where away?" promptly responded the officer of the deck.

"Four points off the lee bow! Blows sperm-whales! Blows! Blows!" came from aloft.

"How far off?" shouted the captain, roused out of his cabin by the alarm, as his head and shoulders appeared above deck. "Where are they heading?" he continued, as he went up the rigging on all-fours.

"Blows about two miles and a half off, sir," replied Mr. Braxton, the mate, looking off the lee-bow with his glasses, "and coming to windward, I believe."

"Call all hands!" said the captain. "Haul up the mainsail, and back your main-yards. Hurry up there! Get your boats ready, Mr. Braxton!"

At the first alarm the men came swarming up the companionway of the forecastle, divesting themselves of superfluous articles of clothing, and scattering them indiscriminately about the deck. Rolling up their trousers, and girding their loins with their leather belts, taking a double reef until supper-time, they flitted nervously here and there in their bare legs and feet, observing every order with the greatest alacrity, and holding themselves in readiness to go over the side of the vessel at the word of command. There is a certain order, systematic action, or red tape, observed on all first-class whaling-vessels, however imperfectly disciplined some of the boat-crews may be. The captain indicates the boats he wishes to attack the whales; the boat-header (an officer) and the boat-steerer (the harpooner) take their proper positions in the boat, the former at the stern and the latter at the bow, while suspended in the davits. At the proper moment the davit-tackles are run out by men on deck, and the boats drop with a lively splash; the sprightly oarsmen meantime leap the ship's rail, and, swinging themselves down the side of the vessel, tumble promiscuously into the boats just about the time the latter strike the water. Although it may be said that there is a general scramble, there is not the least confusion. Every person and thing has the proper place assigned to it in a whaleboat; the officer has full command, but he is subject to the orders of the captain, who signals his instructions from the ship, usually by means of the light sails. The manner of going on to a whale, the number of men and their positions in the boat, and the kind of instruments and the manner of using them, have been perpetuated in this fishery for more than two centuries.

"Clear away the larboard and bow boats!" shouted the captain. "Get in ahead of the whales, Mr. Braxton, if you can. Here, cook, you and cooper lend a hand there with them davy-taycles. Are you ready? Hoist and swing your boats."

Down went the larboard boat and the bow boat almost simultaneously.

"Shove off! Up sail! Out oars! Pull ahead!" were the orders from Mr. Braxton, the officer of the larboard boat, in rapid succession. "Let's get clear of the ship. Come, bear a hand with that sail, do," he added, coaxingly, with his eye on the third mate's boat. "Don't let 'em get in ahead of us."

"All right, sir; here you go, sheet," replied Vera, the harpooner, a well-developed and intelligent American-Portuguese, with his accustomed good spirits.

Hastily laying aside his paddle, like a tiger couchant, with eager eyes upon his prey, he picked up his harpoon, and stood erect, his tall, muscular frame swaying above the head of the boat. He placed his thigh in the clumsy-cleat,—a contrivance to steady the harpooner against the motions of the waves,—and with his long, springy arms turned and balanced the harpoon-pole previous to poising the instrument in the air. . . . Under the motive power of sail and paddle the space between the boat and whale was rapidly diminishing, and apparently they would soon come into collision. The enormous head of the cetacean, as it plowed a wide furrow in the ocean, and the tall column of vapor rising from the blow-holes, as it spouted ten or twelve feet in the air, were to be seen right ahead; the expired air, as it rushed like steam from a valve, could be heard near by; the bunch of the neck and the hump were plainly visible as they rose and fell with the swell of the waves; and the terrible commotion of the troubled waters, fanned by the gigantic flukes, left a swath of foaming and dancing waves clearly outlined upon the surface of the sea. . . .

Mr. Braxton laid the boat off gracefully to starboard, and the mastodonic head of a genuine spermaceti whale loomed up on our port bow. The junk was seamed and scarred with many a wound received in fierce and angry struggles for supremacy with individuals of its own species, or perhaps with the kraken; the foaming waters ran up and down the great shining black head, exposing from time to time the long, rakish under-jaw; but what small eyes!

"Now!" shouted the officer, as if Vera was a half-mile off, instead of about twenty-five feet. "Give him some, boy! Give him—!" But his well-trained and faithful harpooner had already darted the harpoon into the glistening black skin just abaft the fin; the boat was enveloped in a foam-cloud—the "white water" of the whalemen, stirred up by the tremendous flukes of the whale.

"Stern all!" shouted the officer; and the boat was quickly propelled backward by the oarsmen, to clear it from the whale. "Are you fast, boy?"

"Fast iron in, sir; can't tell second," replied Vera; but the zip-zip-zip of the line as it fairly leaped from the tub and went spinning round the loggerhead and through the chocks, sending up a cloud of smoke produced by friction, indicated the presence of healthy game.

"Wet line! wet line!" shouted Mr. Braxton, as he went forward to kill the whale, and Vera came aft to steer the boat, unstepping the mast on his way; for all whales are now struck under sail. The whale, however, soon turned flukes, and went head first to the depths below. Meantime, the other whales had taken the alarm, and with their noses in the air, were showing a "clean pair of heels" to windward.

The boat lay by awaiting the "rising" of the cetacean. Twenty minutes passed, twenty-five, stroke-oarsman began to feel hungry; thirty, thirty-five, and still the line was either slowly running out or taut; but soon it began to slacken. "Haul line! haul line!" said the officer, peering into the water. "He's stopped." The line was retrieved as fast as possible and carefully laid in loose coils on the after platform. "Haul line, he's coming! Coil line clear. Vera!" said Mr. Braxton, shading his eyes with his hand and looking over the gunwale at an immense opaque spot beginning to outline itself in the depths below.

FAST TO A WHALE.

"Look out! Here he comes! Stern all! Look out for whale!"

But the mate's injunctions were received too late. The whale, fairly out of breath, came up with a bound and a puff, scattering the water in all directions, and catching the keel of the boat on the bunch of its neck. The boat bounded from this part of the whale's anatomy to the hump, and, careening to starboard, shot the crew first on the whale's side and then into the water. The stroke-oarsman now began to feel wet. The whale, terrified beyond measure by the tickling sensation of the little thirty-foot boat creeping down its back, caught the frail cedar craft on one corner of its flukes, and tossed it gracefully, but perhaps not intentionally, into the air, as one would play with a light rubber ball. As the boat descended, with one tremendous "side wipe" of the mighty caudal fin, and with a terrible crash that was heard on the ship nearly two miles away, the whale smashed it into kindling-wood.

A WHALE-BOAT CUT IN TWO.

This is only one of the exciting tales Mr. Brown has to tell, and the history of whaling in every country could add many more. He tells us that approaching a whale at all times is like going into battle, and says that many of the deeds remembered by old hands were purely heroic, since the danger might have been avoided by declining to attack the animal under the especially hazardous conditions that often present themselves.

The persecution suffered by whales of all kinds in all parts of the world made the more valuable kinds so scarce by the middle of the present century that many voyages were almost fruitless, not only by reason of small catches, but because the substitutes invented for whalebone, and the constantly increasing use of mineral oils had lowered prices to an almost ruinous level. The American fleets suffered with the rest, until during the Civil War they were nearly swept from the seas by the ravages of the *Shenandoah* and other Confederate privateers.

Since then there has been only a partial revival, accompanied by a good many changes. A few Scotch and German whalers still go to the northern seas, working in the ice, and some American vessels from the Eastern States, and a greater number from California search the Pacific and the waters off Alaska. All or nearly all of these whalers are provided with steam-propellers, having an arrangement by which they can lift the screw out of water and use their sails for ordinary purposes. Many of them chase with a steam-launch instead of the old-fashioned whaleboats, and save their men the back-straining labor of towing a prize perhaps two or three miles to the ship. In place of the hand harpoon they have several forms of swivel-guns and shoulder-guns discharging harpoons and explosive darts by gunpowder, so that a large share of the danger as well as the labor is saved to modern whalemen, who are also much better housed and fed in their large iron steamships than those used to be who wrestled with scurvy in the grim old hulks of half a century ago.

The ships that go up through Davis Straits now frequently winter there, in order to be on hand in May to meet the whales that appear in the first open water, to which the men drag their boats over the ice between their ships and the first open channels. For the same purpose many vessels of the American fleet are accustomed to pass the winter in company under the shelter of islands near the mouth of the Mackenzie River. Here they have a rendezvous where buildings have been erected and means for social comfort have been established, such as billiard tables, books, etc. These western vessels do not force their way into and through the ice, as do those among the eastern archipelagoes, but operate in comparatively open water, as long as it lasts, along the edge of the paleocrystic ice. Delaying the departure of those who mean to return to the Pacific and home until the last moment, it occasionally happens that some are caught and frozen in. These are usually destroyed, but thus far their crews have managed to escape either to more fortunate vessels or to the shore, where, at Point Barrow, the government has built and keeps furnished a strong house, with stores, fuel, and provisions, as a refuge for shipwrecked mariners.

Walrus-hunting is not much followed nowadays by civilized seamen, though the animal is still of great value to the Eskimo and Siberians. It has become very scarce in easily accessible waters, but is occasionally taken by whalers, who find a market for the ivory of its tusks.

Sealing is an industry which still claims considerable attention from the Scandinavians and Scotchmen who go to the coasts and waters about Spitzbergen, Jan Mayen, and Greenland, as well as to nearer resorts, in pursuit of several species yielding oil and valuable hides; and in the North Pacific the pursuit of the fur seal still occupies many small vessels, but seems likely to come soon to an end. Antarctic seals are practically extinct.

The industry of fishing is probably one of the oldest in the world, and it remains among the most important, for the fisheries not only furnish a vast amount of nutritious and pleasant, yet remarkably cheap, food, but many other things useful to mankind. Hence it is not strange to find that in all the early reports of the discovery of new lands and waters that followed one another so rapidly from the fourteenth to the eighteenth centuries, the fish and other sea-animals to be found were always given a prominent place in the list of valuable assets pertaining to each locality. Even the Spaniards and Portuguese, in their insane rush for gold and silver, to the neglect and ruin of everything else, had to pay some little attention to fishing and allied industries in both the East and West Indies; while in the case of the exploitations of new regions by the calmer, more prudent people of western Europe — the British, French, Dutch and Scandinavians,— the value of the harvest of the sea was really more in view, at first, than that of the land, at least when they began to visit and colonize North America. Take, as an example, the history of St. Pierre, Miquelon, and the others that form a group of islets in the Gulf of Newfoundland, half way between Prince Edward Island and Newfoundland. Mr. S. G. W. Benjamin, in whose " Cruise of the *Alice May*" you may find many interesting and picturesque materials for an account of them, tells us a French settlement was begun on St. Pierre as early as 1604, and that tradition says the islands were resorted to by the Basques two centuries before that, as is very likely true.

In 1713 the colony numbered three thousand souls, and had become a very important fishing port. In that very year St. Pierre was ceded to Great Britain, together with Newfoundland, the French being merely allowed permission to dry their fish on the adjacent shores. But when the victory of Wolfe resulted in the loss of Canada to France, she was once more awarded this little group of isles lying off Fortune Bay, to serve as a depot for her fishermen. The French now gave themselves in earnest to developing the cod-fisheries, determined, apparently, that what they had lost on land should be made up by the sea. In twelve years the average exportation of fish amounted to six thousand quintals, giving employment to over two hundred smacks, sailed by eight thousand seamen. The English recaptured the isles in 1778, destroyed all the stages and store-houses, and forced the inhabitants to go into exile. The peace of Versailles restored St. Pierre to France in 1783, and the fugitives returned to the island at the royal expense. The fisheries now became more prosperous than ever, when the war of '93 once more brought the English fleets to St. Pierre. Again the inhabitants were forced to fly. By the peace of Amiens, in 1802, France regained possession of this singularly evanescent possession, and lost it the following year, when the town was destroyed. In 1816 St. Pierre and Miquelon were finally re-ceded to France, in whose power they have ever since remained.

CURING FISH AT ST. PIERRE.

As these islands were of no use to any one for any other purpose, all this
struggle for their possession was in order to retain the privilege and naval
control of fishing in those waters. The French government has carefully
fostered this interest ever since, and now the islands not only have a settled
population of several thousand, but at the height of the season sometimes
as many as ten thousand strangers (sailors and fishermen) congregate at

the principal port, St. Pierre, which is one of the most important centers in the world for the marketing, curing, and export of sea-caught fish.

Of all waters those of the North Atlantic seem to excel in useful fishes; from the oil-shark hand-lining off the coast of Lapland, or the sardine-catching of Spain, to Yankee sword-fishing, this ocean is alive with fish and fishermen, on both sides and at all seasons of the year.

The whole coast of Norway supports this industry, especially around the far northern Lafoden Islands. The North Sea, shallow and cold, is the home of many valuable species that are sought by extensive fleets from Denmark, Holland, and the north of France, while thousands of British sailors make a living along their own eastern coasts and among the islands north of Scotland; but the waters on all sides of the British Isles are fishing waters, especially the English and Irish channels and the western lochs of Scotland; the herring-catch alone is worth eight and a half millions of dollars a year, while Great Britain's mackerel-catch amounts to two millions, and her share of the codfishery to another two millions. Nearly half of all the products of British fisheries are obtained by the use of the beam-trawl — a huge dredge-like bag-net, handled and towed by steamers in pretty deep water, which scoops in everything near the bottom, where the most desirable sea-fishes stay. Among the prizes are the turbot and sole — toothsome and valuable species not known along American shores.

More southerly are the profitable fisheries for pilchards, sprats, and especially sardines — little fishes taken in vast numbers and canned or preserved in various ways. The abundance of sardines, a recent writer tells us, may be inferred from the fact that the Spanish fishermen take annually about one hundred thousand tons of these little fishes, having a value of from $400,000 to $600,000. A peculiar method of capturing the sardines at night prevails in the Adriatic. The location of the shoals of fish is literally felt out by a light sounding-line, and by means of the attraction of a fire of resinous pine the fish are slowly coaxed into some creek or estuary and surrounded with a seine. The demand for wood for use in this and other night fisheries causes a serious drain on the neighboring pine-forests.

The *great* fishery of the Mediterranean, however, is that for tunnies — huge fishes allied to mackerel, sometimes weighing several hundredweight, and regarded in America as poor food. They have been taken by means of pounds and strong enclosing nets ever since classical antiquity, and preserved tunny flesh is still popular in Spain, Italy, and North Africa, while the same fish is the object of one of the principal sea-industries of Japan.

But important as are the catching, preserving, and utilization of these and many other European fishes, they are far outranked by the marine fisheries for the cod and its relatives, the halibut, haddock, hake, etc., in waters about Newfoundland, Labrador, and Iceland, where also great quantities of mackerel, herring, and other food fishes are regularly obtained.

HAND-LINE FISHING ON THE GRAND BANKS.

The principal grounds are on the Banks of Newfoundland, which have been resorted to for more than three hundred years by men from both continents.

The Banks of Newfoundland are a series of shoals — submerged islands, in fact — which lie off the northeastern coast of America from Cape Cod to the farther end of Newfoundland. The shallowness of the water over them

makes them advantageous places for fishing, because many of the species
caught remain near the bottom, and in deep water are therefore beyond
convenient reach. It is possible, also, to anchor there — often a necessity.

But just here are presented some of the worst perils to which fishermen are exposed. Nowhere are old ocean's storms worse than on these
Banks, where the sand is sometimes stirred five hundred feet below the surface. The best fishing comes in winter — the season of the heaviest gales.
The vessels must anchor close together, too, for the areas of good fishing
are small, and if one breaks its hawser, or the anchor drags, there is great
danger of drifting afoul of some neighbor, which is likely to end in the destruction of both. Then there is ever present the danger, in these latitudes
of almost ceaseless fog, of being run down by the transatlantic steamers, in
whose track the fishing fleets must anchor. The skipper keeps his bell
tolling, or a great horn blowing, but if a steamer comes down the wind her
lookout will hardly be able to hear it before it is too late to stop or change
the course of the monster rushing at full speed through the thickness of
mist and flying spray. " Before anything can be done the relentless iron
prow cuts into the schooner, which for a moment quivers and then disappears
into the depths. . . . One of these great iron ships might cut the bows off
a fishing schooner of sixty or eighty tons and not, perhaps, experience a
sufficient shock to alarm the passengers sleeping calmly in their staterooms."

The vessels which go upon this perilous quest are the stanchest, swiftest,
and withal handsomest little vessels that sail our seas. Their rig is adapted
to this purpose, and spreads almost as much canvas as a racing-yacht,
which, in fact, on this side of the Atlantic has been modeled from Banks
fishermen. The best of them probably are those hailing from Gloucester,
Mass., and these are never used for any other purpose.

The old-fashioned hand-line fishing, such as still holds a place in the
mackerel fisheries — although even there it has given way in most vessels to
purse-netting, — is no longer practised in the American codfishery, which
now uses the trawl-line altogether, by which the men have added to the
hardship and danger of their adventurous life as well as to its profits.

This trawl is not a huge dredge as is the beam-trawl of the North Sea
fishermen, from which it has unfortunately copied its name, but is a strong
rope between three and four hundred feet long, having at each end an
anchor and a flag-buoy. It is so arranged that when it is stretched out and
anchored the line will be several fathoms beneath the surface. To this line,
at intervals of six feet or so, are hung short lines, each carrying a stout
hook. When the fishing-ground has been reached, the captain anchors his
vessel, or, if the weather permits, he sails gently to and fro. Previously, six

trawls have been baited with clams brought from home, and one put in each of the six small boats which the vessel carries. Two men now put off in each of these boats and anchor the trawls at convenient distances from each other, in such a way that the trawl-line, with its fringe of hooks, shall be

A FISHING SCHOONER "HOVE TO" IN A GALE ON THE BANKS.

stretched taut and at the proper depth. How long they stay down depends on the weather — five or six hours, or from evening until morning, is the usual period. Then the men go out, and taking up the anchors at one end, haul each trawl into the boat, coiling it in the bottom and taking off the hooks each captive fish as fast as they come to it.

Simple as this sounds, it is terribly hard work. The trawls are heavy and stiff, and armed with dangerously sharp hooks. The busiest season is midwinter, and no dread of cold or danger must stop the fisherman, who boldly ventures in his little dory into the teeth of a howling snow-storm and fast increasing gale, piling the water "mountain-high" about him and encasing his body in a sheet of icy spray; this must he do, in spite of dis-comfort and the imminent risk of death, if he would save from destruction his valuable trawls and the booty they may have hooked for him. A fine day on the Banks of Newfoundland is a rare thing; fog and snow and icy gales are the rule, and only the boldest courage, endurance, and skill will enable a man to resist that ocean and wrest from it his self-support. A vivid picture of the hardships and dangers of fishing on the Banks is to be found in Rudyard Kipling's story, "Captains Courageous."

The intrepid and skilful voyages of our whalers and fishermen, daring every fatigue and danger in the open sea, have been schools for the best seamen of the world. Every nation is glad to draw these sailors into their navies, and it is they who make the bravest yet most cautious captains of our merchant marine, showing to their comrades and to landsmen splendid examples of heroism and fortitude. *This* is the schooling I meant when I said that in its industries we get not only food, but formation of character, from old Ocean,— and this is the highest result attainable from either land or sea.

CHAPTER XII

THE PLANTS OF THE SEA AND THEIR USES

HE ocean was the home of the first living thing, either plant or animal, that appeared on our planet; seaweeds and salt-water animals are found in much older rocks than any that contain the fossils of land life. Moreover, though called a "wide waste of waters," and seeming a complete desert as we gaze upon its restless surface on a dull morning, there is a greater number of animals and plants by count, and quite as large a variety, under the waves as above them, and the bottom of the sea — at all events near its margin — is more populous than any bit of woods you ever saw.

There exists in our ponds and ditches a race of plants so minute that it requires a powerful microscope to examine them. Under this instrument it is seen that they have delicate, flinty shells or armor, which is of a great variety of forms,— coiled, globular, boat-shaped, spindle-like, and so on,— and always beautifully sculptured. These minute and beautiful diatoms, as they are called, move about freely, and were long supposed to be animals: now they are known to be the simplest of seaweeds, consisting of only one cell. Since life first began, these diatoms, and other microscopic plants much like them, have swarmed not only in the fresh waters, but in all the oceans of the globe, furnishing food for mollusks and all the lowly animals whose food is brought into their mouths by the currents. Innumerable, and as wide-spread as the salt water itself, every one of these myriads of minute plants has left a record; for its delicate, glass-like shell was indestructible, and when the bit of life was lost, it sank slowly down to the bottom. What effect toward perceptible sediment could come from a thing so small that it would scarcely be felt in your eye? One or two, or even a million, would go for little; but century after century, through ages too long for us to comprehend, a steady rain of these exquisitely engraved particles of flint showered down upon the still sea-floor, almost as thickly as you have seen motes in a sunbeam, until there was deposited a layer, many feet in thickness, of nothing but diatom-skeletons. Though this went on to a greater or less extent everywhere in the sea, such deposits are not now to be discovered everywhere, because disturbing causes swept the shells away, or broke up the floor after it had been laid down; but in various parts of the world to-day, you may find wide beds of rock made up wholly of such skeletons, soldered together into hard stone; while in some regions the mud of our sea-bottom appears to consist of almost nothing else. The mighty chalk cliffs of Great Britain and the French coast were built up in precisely this way at the bottom of an ancient sea, whence they have been lifted, but they are composed of much besides diatoms.

From the simplicity of diatoms the vegetation of the sea can be traced upward through larger and more complicated kinds of plants until we reach the enormous algæ that break the gloom of black headlands by their brilliant tints, and furnish a lurking-place under their wide-spreading and dense foliage for hosts of marine animals — some hiding for safety, others to watch for prey.

Seaweeds grow in all latitudes, even close to the pole, but mainly along the shore, for below the depth of about one hundred fathoms none but microscopic forms are known. These latter float about, of course, and many of them have been thought to be animals because they seem able to move at their own will. They come to the surface as well as haunt the depths; and the Red Sea takes its name from the fact that a minute carmine-tinted alga occasionally rises to the surface in throngs so dense and wide as to tinge the water for miles at a stretch. The same thing occurs in the Pacific, where the sailors call it "sea-sawdust."

The proper home of the seaweed, however, is a rocky shore between tide-marks or just below them, and it is because the eastern coast of the United States is deficient in rocks — at least south of Cape Cod — that this is poor in algæ, compared with other regions. The seaweed has no roots, and only clings to the rock for support; shifting sand therefore would not hold it, and there are great sandy deserts under the ocean, bare of algæ, as some land regions are sandy deserts naked of terrestrial plants.

It often happens, however, that masses of weed will be torn away from their moorings and set adrift. This does not necessarily kill them, for they go on flourishing while afloat, and such is supposed to be the origin of those great areas of "gulfweed" vegetation in mid-ocean called "sargasso seas." You will remember that a branch of the Gulf Stream, striking over toward the Moorish coast of Africa, is turned southward there, and sweeps down to the equator, then westward again, circumscribing a broad region in the middle Atlantic whose only currents go round and round in a slow whirlpool; and here it is that the gulfweed concentrates in masses sometimes dense enough to impede the progress of a ship—Columbus reported among the wonders of his first voyage the trouble he had in sailing through it — and covering an area between the Azores and the Bahamas as large as the Mississippi valley. This is the Sargasso Sea ordinarily referred to in

THE MARBLED ANGLER ON ITS GULFWEED RAFT.

books, but it is not the only one. A thousand miles west of San Francisco there is a similar collection of floating plants, and others exist under like conditions in the southern oceans.

These floating meadows, as it were, are chosen as the abode of a long list of animals that rarely quit the safety and plenty of their precincts.

Among these are innumerable pretty jelly-fishes, sea-worms, and mollusks without shells, which cling to the buoyant plants, and perhaps feed solely upon them. Here are to be had in abundance the fairy-like, rare pteropods, the richly purple janthinas towing their curious rafts of eggs, and no end of small crabs. Here a small fish, something like a perch, spends his whole time building a nest like a bird's in the tangled weed-masses, and carefully guarding his treasures against the large marauding fishes that haunt the place to the dread of its peaceful inhabitants; and here those far-flying birds, the wandering albatross and the petrels, hover about in search of something to capture and eat. The Sargasso Sea is an extremely interesting part of the ocean, except to the luckless sailor becalmed and balked in its midst, as was Sir John Hawkins when he penned the following quaint observations, some three centuries ago :

Were it not for the Moving of the Sea, by the Force of Winds, Tides and Currents, it would corrupt all the World. The Experience of which I Saw *Anno* 1590, lying with a Fleet about the Islands of Azores, almost Six Months, the greatest Part of the time we were becalmed, with which all the Sea became so replenished with several sorts of Gellies and Forms of Serpents, Adders and Snakes, as seem'd Wonderful : some green, some black, some yellow, some white, some of divers Colours, many of them had Life, and some there were a Yard & a half, & some two Yards long; which had I not seen, I could hardly have believed.

In favorable places a surprising variety of seaweeds can be picked out, and books exist by which you may learn the method of classification and names of the different species, the chief of which, for America, is Harvey's splendid work, published by the Smithsonian Institution. Not only in the shape and colors of the *fronds* (as the leaf-like expansions or branching tufts of the stem are called) do seaweeds differ greatly among themselves, but in size, varying from many

A PIECE OF GULFWELD.

It is inhabited by two sea-slugs, protected by their resemblance to it, leaflets, and by small crustaceans, hydroids, etc.

diminutive or even microscopic sorts to the cable-like growths of California, which would measure a quarter of a mile in length if stretched out.

Algæ, as I have said, constitute, with very few exceptions, the whole vegetation of the salt water, together with a large part of the vegetation in fresh water; and they serve the same useful purpose there that land-plants do for the dry parts of the globe, continually making and throwing off the oxygen which is necessary to keep the water as well as the air pure. To this end they do a very important work.

This is not the whole of their service in ocean matters, however. I think it may be said that if it were not for seaweeds animals could not live in the ocean, as truthfully as that if it were not for herbage no animals would be able to exist on land. Seaweeds are fed upon directly by all sorts of salt-water life, from mollusks as big as your thumb to turtles the size of a dining-table, and they make a shelter for thousands of little fellows who never leave their shadow.

But this is a small part of the story. The diatoms, and other minute plants like them, form the main portion, if not all, of the food of a large number of sponges, polyps, mollusks, and other stationary, sluggish creatures, that otherwise, so far as I see, would not be able to live at all. These, in turn, are fed upon by larger predaceous animals. Thus, though the fishes and cetaceans may never bite a seaweed themselves (those large marine herbivores, the manatee and dugongs, subsist almost wholly upon it, however), they depend for food upon creatures that do. We may say, therefore, that the algæ form the basis of all ocean life.

Men have been able to make marine plants of service to them also — a resource more important formerly than now. In the last century, for example, the kelp trade was the one great industry of the islands at the west of Ireland and Scotland, employing thousands of persons, and paying vast revenues to the lordly owners of the shores. Kelp is the name of any large, leathery sort of seaweed, whose leaves float at or near the surface, supported by bladder-like expansions; but in this case the word meant the ashes of any seaweed dried in the sun and then slowly burned in kilns, clouding the air with huge volumes of strongly odorous smoke. The slow burning of the seaweed left the ashes fused into a solid mass, which was broken up like stone before being sold. In France this substance was called *varec*; and in Spain, where the algæ were mixed with beach-plants, cultivated for the purpose, and burned in shallow pits in the ground, it went to market as *barilla*.

In those days, kelp ash was the only source of the valuable alkali soda needed in manufacturing glass and soap. Then a French chemist discovered how to make such soda out of common salt, and the kelp ovens were abandoned, except a few in Scotland, supplying the demand for iodine and several other chemicals contained in this residuum which is so rich in iodine, used in photography and in medicine, that a ton of kelp ash will sometimes yield twenty pounds; yet only about 100,000 pounds are now produced in this way, while five times as much is obtained by chemical treatment of Chile saltpeter. It is a curious fact that barbarous people have long chewed seaweeds as a remedy in diseases for which physicians now prescribe iodine. Iodine is a violet dye, and the bluish and purple tints of many algæ, shells, and sea-animals appear to be due to the large amount of this element in sea-water.

Seaweeds and other marine plants, like eel-grass, are collected in great quantities by farmers in all parts of the world to be used as a fertilizer.

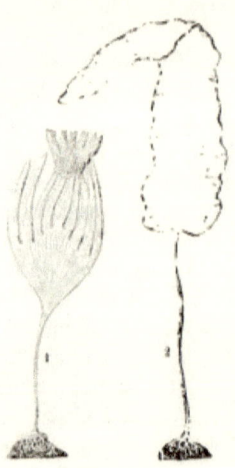

SEAWEEDS.

a. Laminaria digitata, b. L. longicruris

Shell-mud, dead fish, and other marine products are also of high value as manure, on account of the large proportion of lime, carbon, and soda which they contain. Indeed, there is a kind of seaweed growing at great depths called the nullipore, which takes up so much lime from the water that its substance becomes almost like stone, so that the plant retains its shape and full size when dried. Some of these nullipores are beautifully fan-shaped, scarlet or pink, and are often seen in museums, marked *corallines*.

To return to the gathering of seaweeds by farmers, nowhere is it more customary than in some parts of New England. Thus the well-known Second Beach, just east of Newport, is in the fall of the year the scene of a vast activity in this direction. "It may easily happen," we are told, "that the pilgrim to Whitehall, topping the hill on a brilliant autumn morning, shall come upon a scene in which quiet plays no part. The seaweed, that harvest which, ripening without labor, is neither bought nor sold, is setting inshore under the urgings of wind and tide, and scores of farmers have crowded to the spot to gather it. An artist could hardly wish a better subject for his pencil than one of these wild harvestings — the plunging horses, forced far out into the surf, their slow return, half swimming, half wading, dragging the heavily loaded rakes which leave behind them a long furrow of foam, the heaped-up kelp glistening in the sunshine, the oxen, yoked by fours, waiting for their load, the shouts of the men, the dash, the excitement, and beyond and above all, the wonderful blues and iridescent greens which are the peculiar property of Newport waters and the Newport sky."

Cattle and horses that are accustomed to rough pastures, like the Scotch and Irish moors, eat seaweed and thrive on it, especially as winter fodder, and from several species are derived dishes for our own tables. The Irish moss, or carrageen,— which is not a moss at all, but a seaweed,— is the most important of these, and grows on both sides of the northern Atlantic. In England the market supply comes chiefly from the western coast of Ireland, while Massachusetts Bay gives America all that is wanted, principally the red, coral-like *Chondrus crispus.* The little port of Scituate, Massachusetts, is the chief point of supply, where many thousands of pounds are gathered. In early June, two or three hundred men and women go to the rocks at low tide and pick off the small brown plants, each man getting about a barrel in one day's work. When the tide rises, the people get into small boats and pull up the moss with rakes.

The moss gathered each day is taken to the beach, where a gravelly space has been prepared, and is spread out to lie bleaching during all of the next day, when it is taken up, washed in tubs, and again spread out. The washing and drying in the sun continue for seven days, by which time it has bleached to a yellowish white. In cookery, jellies, *blanc mange,* and various methods of boiling in milk and mixing in soups are used to make it palatable. Besides being of value for food, carrageen serves to make sizing used by paper-makers, cloth-printers, hatters, and so on, to clarify beer in the brewery vats, as a medicine, and to make bandoline for stiffening the hair.

Other species beside the Irish moss serve as food in Europe, generally in a raw state, often proving the only salty relish which the Irish peasant has to eat with his potatoes. One of these is the *dulse* of the Scotch (the *dillisk* of Ireland), which also abounds in the Mediterranean, and is there made into a soup. The natives of the South Sea Islands eat algæ, which are extraordinarily abundant and varied in Oriental latitudes; and the poor among the Japanese and in the interior of China, where the weed is sent dried, prize it especially, because it has a sea flavor and saves salt, which with them is a costly luxury. These people mix it with vegetables and other materials, to form thick, delicious soups and dressings. A peculiarly bad-smelling sauce, prepared from seaweed, is among the exports China sends to Europe as a condiment.

Along the shores from Japan to Sumatra grows an alga which the natives of those coasts dry and keep as long as they please. When the substance is wanted they steep some of the dried pieces in hot water, where the weed dissolves, and then, having been taken off the fire, stiffens into a glue which is said to be the strongest cement in the world.

A kind of false isinglass, also, is a product of the Eastern seaweeds, and it not only enters into the pastry and confectionery of Chinese bakers, but serves to varnish and glue thin paper and to stiffen the light transparent gauzes of fine silk used in making Oriental screens, fans, hangings, etc., so that painters can decorate them. With a poorer quality the bamboo stretchers of paper umbrellas, lanterns, and various toys are smeared to give them hard and polished surfaces.

Seaweed has also been used in the manufacture of paper, and its complete success in this branch of industry is as yet hindered only by the difficulty of perfect bleaching. Certain species of it are utilized in enormous quantities by upholsterers as stuffing for sofas, chairs, and mattresses; in Japan it is formed into a substitute for window-glass; ornaments and small articles of use, like knife-handles, are made by several nations out of large dried seaweeds; and, finally, albums of preserved fronds are one of the prettiest things to be found in a naturalist's cabinet.

The great majority of seaweeds grow between tide-marks, and they undoubtedly perform an important service in preventing the wear and tear of the coast in many situations. Some, however, grow in much deeper waters, and these, also, may serve as breakwaters of no mean strength. Such is the case, for instance, at San Pedro, near Los Angeles, California, where the abundant growth offshore forms such a barrier to the ocean rollers as to turn the open roadstead into a calm harbor within it.

This belongs to the group of gigantic kelps of which those at the Falkland Islands and about Tierra del Fuego are other and noted species. Were it not for the growth of this strong, cable-like, buoyant plant, large numbers of other plants and sea-animals would find it impossible to exist exposed to the violence of the South Pacific waves. Sometimes the stems reach twelve hundred feet in length, and the bladders by which the immense fronds are buoyed up are as big as kegs.

This gigantic seaweed is plentiful all along the Pacific coast of America to Alaska, and the natives of our northwest coast used to make extensive use of it in the way of ropes, etc. It was from this weed that, by a careful preparation, they made the lines for their harpoons and deep-sea fishing; and the bladders furnished them ready-made receptacles for eulachon oil, for water for their seatrips, and for other liquids.

A California correspondent of the New York "Evening Post" gave a pretty picture, not long ago, of one of the kelp patches at St. Nicholas Island, where the beds of this wonderful plant reach out for a mile or more, growing up from the rocks below and forming an effectual break ; the seas losing their force in their effort to pass through the submarine meshwork.

The vines constitute a veritable forest, and, drifting over it in fifty or sixty feet of water, you may see a perfect maze of stems with broad leaves waving gracefully in the current, forming arbors, arches, and colonnades. Here, poised idly, in rich contrast to the olive-hued mass, may be seen fish of a bright golden color, others in tints of blue and green. The sea swell coming in causes an undulatory movement, and the long colonnades seem to melt one into another, reappearing in different shapes. When the leaves reach the surface, the shore wind, sweeping down from the hills, lifts them from the water, and they flutter in the air like mimic sails. Each leaf is a study. Many are encrusted with a delicate bryozoön, which presents the effect of white lace

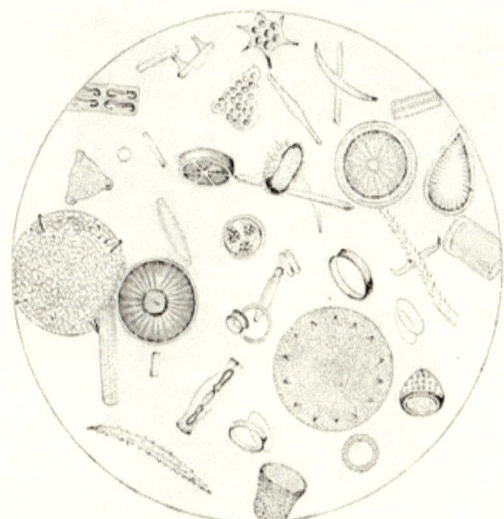

DIATOMS, MAGNIFIED, IN A DROP OF WATER.

upon the surface, while a close inspection will reveal minute anemones, coiled tubular worms, which throw out flower-like organs of exquisite beauty ; while flat shells lie among them, and crawling here and there are marvels of animal life, shell-less mollusks, which so mimic the weed that it is almost impossible to distinguish them.

This protective feature is a characteristic of life among the kelp forests that line the entire Pacific shores of North and South America, many animals simulating it so perfectly in color that the best-trained eyes often fail to observe them. This is especially true of the crabs and shell-less mollusks. The latter have not only assumed the exact tint of the weed, but are often covered with barbels of flesh that simulate the tangles of the substance. Upon the backs of the crabs are singular markings in green and white, which so resemble the minute incrustations of the kelp that the resultant protection is complete. [Compare illustration on page 252.] Each vine is fastened to a stone, and the clinging roots shelter hordes of creatures of various kinds — deep-water crabs, octopods, starfishes, and a host of others.

A MARINE NATURALIST.

CHAPTER XIII

ANIMAL LIFE IN THE SEA

HE primitive idea of the ocean was that it was a vast desert, and a strange disbelief in its being inhabited by more than the very few forms that everybody was compelled to recognize persisted up to quite modern times among those who should have known better. Pliny boldly asserted, for example, that nothing remained in the Mediterranean Sea unknown to him after he had made a list of 176 marine animals! But now we know that the sea teems with living beings as densely as do the fresh waters or the air. In it began the life of the globe, for the fossil records of the rocks show that the first animals lived in the ocean, and that ages passed before any of them began to people the newly formed lands and breathe the atmosphere instead of the air in the water; and, abundant as oceanic life now is, the paleozoic seas held immensely greater hordes, of which many forms were giants as compared with those of our day. Some of the old straight chambered shells were twelve feet long; and I have seen fossil ammonites, extinct relatives of our coiled pearly nautilus, which when alive must have been too heavy for a man to lift. The fishes, too, could tell great stories of the glory of their ancestors in size and strength and numbers. Some of them wore solid coats of mail upon their heads, and could do battle even with the huge swimming reptiles that were the dreaded tyrants of the Mesozoic deep.

Life in the ocean in those old geologic days was a long guerrilla warfare — every animal guarding against attack, and at the same time watching sharply for an opportunity to seize and prey upon some weaker companion. As for the foraminifers and other microscopic creatures, they were countless, and their skeletons, singly invisible, have by accumulation built up great masses of rock, like the chalk-beds of England and France.

Though lessened in numbers and reduced in size, because the land has gradually won over to its side many sorts of animals which in former ages were exclusively confined to the water, and for other reasons, the sea still holds its share of every "branch" and "class" (except birds, and it may almost claim some of them, such as the albatross, penguins, and petrels), and a majority of the "orders" of animal life. Glance at the catalogue: Foraminifers, sponges, and polyps are chiefly confined to salt water; starfishes, urchins (or sea-eggs), and the like, wholly so: mollusks (next higher) are principally oceanic, and the majority of the crabs inhabit salt water. Among the last-named one species, the common horse-foot (*Limulus*) of our shores, remains as the solitary representative of that immense and varied group, the trilobites, which so crowded the Paleozoic sea-bottom that some rocks — for instance, the limestones of Iowa — are packed almost as full of their fossils as is a raisin-box of raisins.

None of the insects is truly marine, yet some of them are seafaring, truly, for they spend their lives on drifting sea-wrack, or on beaches just out of reach of the tides; but most of the true worms are dwellers in the mud of sea-shores and sea-bottoms. No one knows of any land fishes; but I need not tell you that fishes throng in the fresh waters as well as in the salt, and that many species inhabit both at different seasons.

In respect to the reptiles, of which the ancient oceans contained gigantic and horrid types, I do not know any now that are truly oceanic except the turtles, if you leave out the "sea-serpent," of which we hear so many wonderful and not quite satisfactory tales. You will hear of "sea-snakes" in the East Indies, but they are only certain kinds of serpents which swim well, and pass the most of their time in the salt water, as several species of our own country do in the rivers and ponds; all the oriental sea-snakes are venomous.

It is in this manner, too, that we may count certain birds, such as the petrels, auks, penguins, albatrosses, frigate-birds, and their kin, as belonging to the ocean. They spend all their life flying over the waves, seeking their food there, and some of them rarely go ashore, except to lay their eggs and hatch their young on remote rocks, resting and sleeping on the billows, when not busy at their hunting. In the highest rank of all, however, the mammals, several families are natives of the "great deep" — the whales, dolphins, and porpoises, the seals and walruses, and the manatees and dugongs. But all these must come to the surface to breathe, not having gills like fishes, but true lungs.

As it is only within the last thirty years that machinery suitable for deep-sea dredging has been invented, so it is only lately that we have been able to learn much as to the population of the ocean beneath the surface layer and marginal shallows. Now by means of beam-trawls, dredges, tangle-bars, etc., worked by steam-machinery on shipboard, naturalists may scrape up the bottom-ooze and obtain living objects or their bony relics at the depth of even 3000 to 4000 fathoms or more than four miles, for living beings are found in these profound abysses. Many scientific expeditions, such as those of the English exploring steamer *Challenger*, about 1874,

have carried out these dredging investigations, and the United States Fish Commission possesses the large, specially built, sea-going *Albatross*, provided with all the necessary apparatus for deep-sea exploration. By means of these and other vessels an enormous amount of study —all useful in ascertaining the habits and methods of reproduction of food-fishes—has been carried on by American marine naturalists.

LANDING THE BEAM-TRAWL ON DECK.

It appears that as you go further and further from shore, and into deeper and deeper water, the fewer animals and plants are obtained, and that very few species indeed which live along shore are to be found also at a depth greater than about 100 fathoms.

Almost all animals, moreover, have a limited distribution in the sea, as is the case among those on land, though we cannot always, or perhaps often, say why the limits we find should exist: one sort of crab, or mollusk, or polyp, appearing *here* and another different one exclusively *there*, when the conditions seem to us very similar, and no barrier is perceptible. It is not easy to explain why a certain sort of cowry, for example, should be found only along a particular strip of coast, when nothing that we can see prevents its extending its range much further. It is believed that the *temperature* of the water is the chief fact which sets these invisible boundaries to the wanderings of animals living near the surface, only a few of which are very wide-spread in their distribution. The direction and character of the ocean currents have much to do with the geographic distribution of oceanic life, as has been mentioned in Chapter II (page 25).

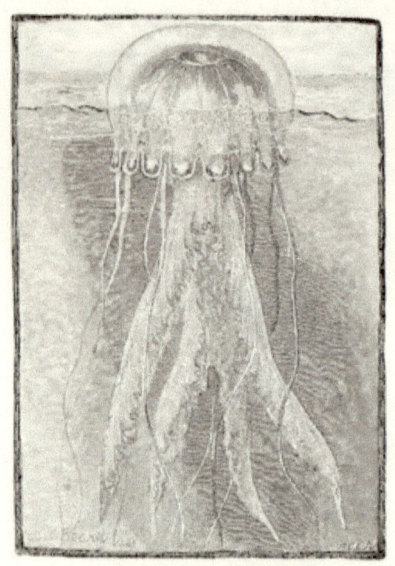

A TYPICAL JELLYFISH.

This species (*Pelagia cyanella*) is a characteristic oceanic discouple sous marine, countien along the Atlantic coast of the United States; it is semi-transparent and lustrous pink.

Now in deep-sea life the case is different. Here temperature cannot be of so much account, since only a short distance down, the water becomes almost as cold as ice, and preserves this uniform chill all around the globe. The life found at a great depth, too, is very wide-spread, instead of restricted in its range, often occurring in two or more ocean basins; but here the restriction is an up-and-down one, rather than horizontal, and the secret is found in the word *pressure*. Few animals are able to live both in the shallows and under the enormous weight of sea water three or four miles deep.

This has recently (1897) been summed up very clearly by Prof. Arthur P. Crouch, in an article in "The Nineteenth Century," from which it will be worth while to quote a paragraph or two:

The conditions under which they [that is, deep-sea animals] have to live in the abysmal areas seem very unfavorable to animal existence. The temperature at the bottom of the ocean is nearly down to freezing-point, and sometimes actually below it. There is a total absence of light, as far as sunlight is concerned, and there is an enormous pressure, reckoned at about one ton to the square inch in every 1000 fathoms, which is 160 times greater than that of the atmosphere we live in. At 2500 fathoms the pressure is thirty times more powerful than the steam pressure of a locomotive when drawing a train.[1] As late as 1880 a leading zoölogist ex-

[1] It does not follow that these creatures are conscious of this pressure, any more than we are of the pressure upon us of the fourteen pounds to the square inch of our atmosphere. The point is that they *do* feel it when they rise upward to a point where the pressure is distinctly less, just as we are conscious of a difference when we ascend in a balloon or climb a very high mountain, and after a time we find that we cannot go any farther. Land animals therefore have a vertical limit to their distribution as well as sea animals, and for analogous reasons.— E. I.

plained the existence of deep-sea animals at such depths by assuming that their bodies were composed of solids and liquids of great density, and contained no air. This, however, is not the case with deep-sea fish, which are provided with air-inflated swimming-bladders. If one of these fish, in full chase after its prey, happens to ascend beyond a certain level, its bladder becomes distended with the decreased pressure, and carries it, in spite of all its efforts, still higher in its course. In fact, members of this unfortunate class are liable to become victims to the unusual accident of falling upwards, and no doubt meet with a violent death soon after leaving their accustomed level, and long before their bodies reach the surface. . . .

The fauna of the deep sea — with a few exceptions hitherto only known as fossils — are new and specially modified forms of families and genera inhabiting shallow waters in modern times, and have been driven down to the depths of the ocean by their more powerful rivals in the battle of life, much as the ancient Britons were compelled to withdraw to the barren and inaccessible fastnesses of Wales. Some of their organs have undergone considerable modification in correspondence to the changed conditions of their new habitats. Thus down to 900 fathoms their eyes have generally become enlarged, to make the best of the faint light which may possibly penetrate there. After 1000 fathoms these organs are either still further enlarged or so greatly reduced that in some species they disappear altogether and are replaced by enormously long feelers. The only light at great depths which would enable large eyes to be of any service is the phosphorescence given out by deep-sea animals. We know that at the surface this light is often very powerful, and Sir Wyville Thomson has recorded one occasion on which the sea at night was "a perfect blaze of phosphorescence, so strong that lights and shadows were thrown

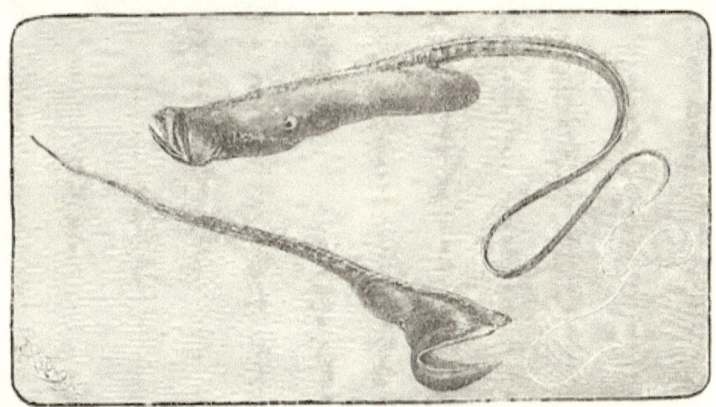

THE BOTTLE-FISH AND THE PELICAN-FISH.

on the sails and it was easy to read the smallest print." It is thought possible by several naturalists that certain portions of the sea bottom may be as brilliantly illumined by this sort of light as the streets of a European city after sunset.

One of the most striking examples of this vertical distribution, which forms layers of animal life, as it were, in the ocean from the abysses to the shallows, is shown by the coral-reefs. The foundations of these polyp-built

AT THE BOTTOM OF THE
TROPICAL SEA.

The large floating object is the phosphorescent, compound, oceanic hydrozoon *Agalma elegans*, a physophore related to the jellyfishes. Its tentacles trail over dead corals,— madrepore, brain-corals, etc.; while the living reef beyond is crowned by branching corals, corallines and seaweeds.

barriers or islands are laid by the millions of minute individuals of one solid, heavy kind of coral which can flourish only in pretty deep water. When these have reached their highest growth they cease to propagate there, and a second kind comes and colonizes upon the summit of this massive foundation and carries the work a little farther up. Then these die off, and a third kind plants itself upon their remains and carries the structure to the top, near the surface of the sea, where many surface-corals, corallines, and various other limy and flinty plants and animals help to erect a dry reef, upon which land vegetation can find a root-hold, and where, after a while, men may dwell. When these coral-built islands are ring-shaped they are called *atolls*, and are believed to be living crowns about the summits of submerged mountains.

Men make use of something in nearly every branch of ocean life, from humblest to highest. The lowest of all, as I have already said, are the foraminifers; it is their skeletons which make up our common chalk. A close ally of theirs is the sponge, of which a dozen or so varieties are sold in the shops. Sponges come chiefly from the Mediterranean, the Persian and Ceylonese waters of the Indian Ocean, and from the Gulf coast of Florida. In the Old World they are obtained chiefly by diving. Men who are trained from boyhood to this work go out to the sponge-ground in boats on fine days. Fastening a netting-bag about their waists, and taking a heavy stone in their hands, they dive head-foremost to the bottom,—often twelve or fifteen fathoms below,—tear the sponges from the rocks, and rise with a bagful, to be dragged almost utterly exhausted into their boat, often fainting immediately after. This requires them to hold their breath under the water for two minutes or more; but none but the most expert can do that, and a diver does not live long. In Florida, however, the sponge-gatherers do not dive, but go in ships to where the sponges grow, and then cruise about in small boats, each of which contains two men: one steers, while the other leans over the side searching the bottom. In order to see it plainly, he has what he calls a "water-glass"—a common wooden pail the bottom of which is glass. Pressing this down into the water a few inches, he thrusts in his face, and can then perceive everything on the bottom with great distinctness. When he sees a sponge he thrusts down a long, stout pole, on the end of which is a double hook, like a small pitchfork, set at right angles to the handle, and drags up the captive.

The sponges, having been obtained, must be put through long operations of rotting, beating, rinsing, drying, and bleaching before their skeletons — the serviceable part — are fit for use. Only a few, however, out of the large number of species of sponges have any commercial value.

The limy skeletons of the coral polyps form what we term "corals." The round white ones and the variously branching ones may come from any one of several parts of the equatorial half of the globe, and are of value chiefly as mantel ornaments. The red coral of which necklaces and other bits of jewelry are made, especially at Naples, is procured by divers about the shores of Sicily and Sardinia, and its gathering, cutting and mounting into ornaments, form a flourishing industry in southern Italy.

Rising in the zoölogical scale to starfishes and sea-urchins, I can only say that the starfishes interest oystermen because they prey upon their oysters, and the former often do enormous damage to planted beds, especially in Long Island Sound. In the old days it was thought that medicines made out of the "stars" and the "sea-eggs" were very potent in certain diseases. The trepang — some one of several sorts of holothurian, an elongated creature related to the starfish, and covered with a prickly, leathery hide, so that it looks like a sort of sea-cucumber — which is dried and eaten by the Chinese and Malayans, belongs here too; considerable quantities of these queer food-creatures are gathered by the Chinese along the coasts of Mexico, Southern California and the outlying islands, and are sold in San Francisco mainly for export to Asia. The sea-urchin itself is eagerly sought as food by the Indians of the American northwest coast.

Coming to crustaceans — do we not eat crabs gladly, from the "shedder" to the huge lobster? On the coast of Maine whole villages of seaside people get their support almost wholly by catching lobsters and canning them to send abroad. In Virginia and North Carolina, at certain seasons, hundreds of men are engaged in catching and shipping crabs for market, and in Louisiana large factories are devoted to canning shrimps, which are also extensively used as food in the Old World, where they are cooked by parching or boiling, and sold by peddlers in the streets.

This brings us to the mollusks, in our glance at the useful animals of the ocean; and to prove *their* importance, it is enough to remind the reader that these include the "shell-fish" of our coasts — the oyster, clam, mussel, scallop, cockle, and all the rest — not a few!

I found by my long study of the subject, when, in 1879 and 1880, I was gathering statistics of the United States shell-fisheries for the United States Fish Commission and the Tenth Census, that at that time there were taken from our waters, of oysters alone, almost 23,000,000 bushels each year, worth to the oystermen about $13,500,000. During the twenty years that have elapsed since that investigation — the figures of which you may obtain in full in my Report to the Tenth Census upon the Oyster Industries — these amounts have largely increased.

This business employs over 100,000 persons in this country alone; and oysters, clams, and other shell-fish are gathered all round the globe, forming one of the most important of all natural supplies of food. In the most thickly populated parts of the world the natural supply of oysters

long ago ceased to suffice for the demand, and artificial propagation and
cultivation were resorted to and now prevail on both sides of the North
Atlantic, and to a less degree elsewhere.

The Romans, away back in the days of Horace, raised oysters in ponds
along the Italian coast, and Eastern nations preserved the custom during
the middle ages, when Europe was doing little except quarreling and
making pretty pictures on parchment. More recently the French of the

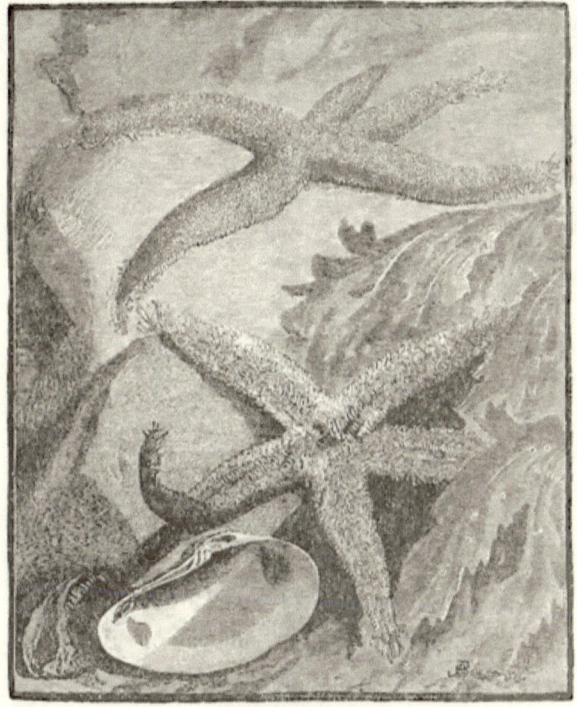

STARFISHES AT HOME.

This is the common eastern American form (*Asterias vulgaris*) upper and under views.

Channel coast took it up, and the English followed, finding that their natural
oyster and mussel beds were becoming exhausted. The same fate has overtaken our oyster-beds everywhere north of the Chesapeake, and largely
there; so that now nearly all the oysters brought to market are those which
have been raised upon private planted beds, which men own or lease and

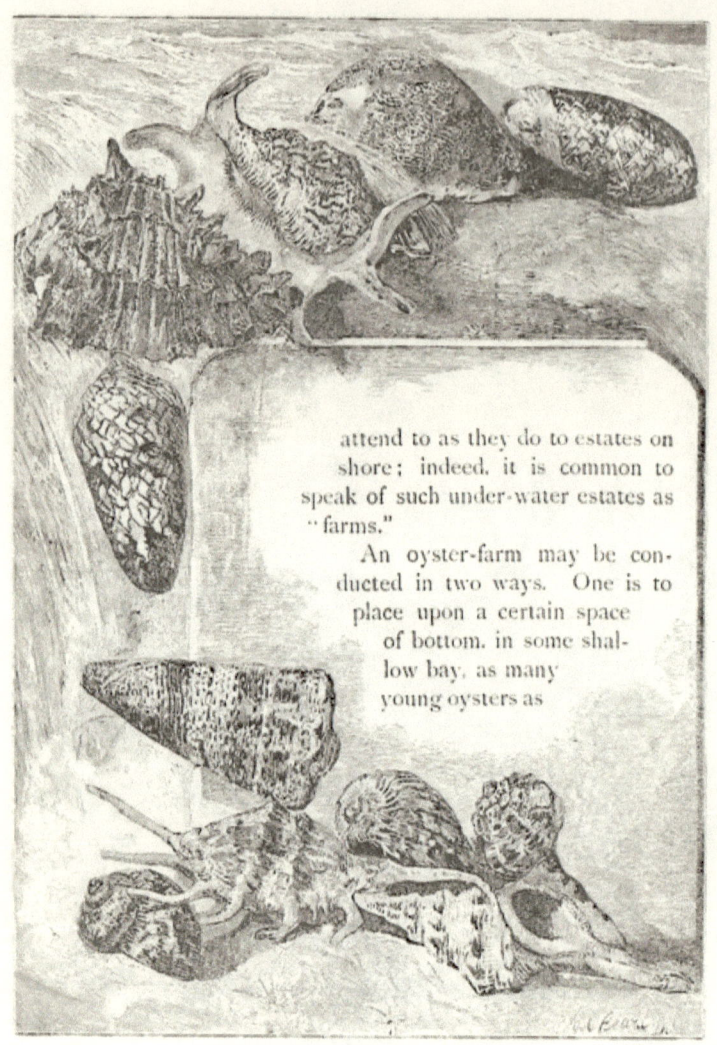

attend to as they do to estates on
shore; indeed, it is common to
speak of such under-water estates as
"farms."

An oyster-farm may be conducted in two ways. One is to
place upon a certain space
of bottom, in some shal-
low bay, as many
young oysters as

SEA-SHELLS IN THE SURF.

it will conveniently hold. These young oysters, generally hardly bigger than your thumb-nail, are dredged in summer from certain reefs in deep water, where the oysters are never allowed to grow to full size; and to a large extent they are brought northward by the ship-load from Maryland and Virginia, which have more "seed," as it is called, than they need for their own planting. These young oysters, protected from harm, and having plenty of space to grow, come to a proper size for market in about three years, and are then gathered by their owners and sold.

Another method is to spread old shells, pebbles, etc., on the bottom, to which the floating eggs emitted by adult oysters in the neighborhood adhere. The thick "catch" of infant mollusks hatched from these captive eggs is then taken up and respread in a more scattered way upon new ground, and is allowed to grow to maturity. The oysters raised by either of these methods are of better appearance and taste, as a rule, than those that grow naturally, because each has room enough to perfect its proportions.

Mussels, clams of many varieties, and even sponges and peak-shells, are also cultivated to some extent, each according to the plan its natural habits make advisable. In this way certain great areas of favorable ocean-bottom have become as valuable as the neighboring shore-land, or even far more so, if you compare, acre for acre, the yield of the crops below with those above the water-line.

But mollusks are useful in many other ways than as human food. As they are known to be the principal food of several valuable fishes, enormous quantities are devoted to baiting hooks in both hand-lining and trawling for cod and similar commercial species. The quaint squids are mollusks, and these are especially useful for bait in certain places and seasons, and are taken in the North Atlantic in vast numbers for that purpose.

The shells of mollusks are applied to a surprising variety of purposes, from paving roads to making shirt-studs, while their natural beauty has suggested their utilization as ornaments in a hundred ways. We cut them up by the million into buttons and various small objects, such as parasol handles, and polish and fashion them into all sorts of knickknacks, thus giving employment to thousands of persons. Many ship-loads of shells are brought to New York from the West Indies every year for such purposes. I need not dwell upon this, but turn to the interesting subject of pearls.

Mother-of-pearl is the bright inside surface, or *nacre*, of the large oyster that gives us pearls, which are themselves composed of the same substance formed in a nodule around some intruding substance, like a grain of sand, which irritates the mollusk's skin until it is made smooth and comfortable by this iridescent coating.

MELEAGRINA.

Meleagrina (Avicula)
margaritifera.
b byssal foramen or
notch; g. suspen-
sors of the gills.

Bivalves yielding this beautiful substance exist in various parts of the world; but in America the only fishery for the pearl-oyster is in the Gulf of California, and that is by no means as productive as it used to be. The season for pearl-fishing on the Pacific coast of Mexico is from June to December, but the diving can be done only in good weather, and for about three hours at the time of low water, since the tide there

CASSIDIDÆ.

Helmet-shell (*Cassis*
flammea).

rises twenty feet, which would make a large dive of itself; and, besides, the currents are troublesome during high water.

At the right hour the Mexicans go out in their canoes, one man of the four or five in each canoe paddling, while the rest scrutinize the bottom. It may be rocky and weed-grown, but the water is clear, and their practised eyes detect a single round oyster where you or I certainly would overlook a dozen of them. Then down a man goes and brings up his prize, with perhaps some additional ones. Sixty or eighty feet is not too deep for these adventurous divers, who will stay a whole minute upon the bottom. No food is eaten by these men on the day they dive until their labor has been done.

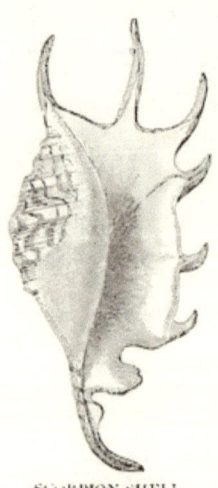

SCORPION-SHELL.

(*Pterocera lambis*.)

Western Australia is another fruitful field for pearl-oysters, and until a few years ago they were taken there by native blackfellows, diving without weights or any other assistance in any water not more than ten fathoms deep. The inshore shallows have now been so cleared of shells that the only profitable industry is to go down in deep water in diving-dress and make a thorough clean-up of each "patch" where the shells seem numerous.

The divers find it an interesting and curious world where they work, but one full of fright and peril. Some men who attempt it are so unnerved that they will never make a second

MITER-SHELLS.

a. Mitra vulpecula. b. Mitra
episcopalis.

descent. None can endure the practice long without ill health resulting; and the native Australians will never enter a diver's dress, declining to go down where it is too deep to dive naked.

As for the dangers, drowning by some accident to the apparatus, or through the stupidity of the boatmen above, is only one of them. The warm waters in which these men work are the home of the

VENUS' COMB, ONE OF THE MUKICES OF CHINA.

largest and most deadly sharks, and of various other submarine creatures one would rather not meet in their own element. Of them all the sharks are most to be dreaded, especially by the naked men. As a rule, however, they are easily frightened away, or can be avoided by the clever swimmer, who quickly stirs up the mud of the bottom, and rises in the fog before the dull shark discovers that he has gone. East Indians are said to fight sharks quite fearlessly, stabbing them with a knife as they roll over preparatory to a close attack. I have read a story to the effect that formerly the Mexican Indian divers on our western coast used to take down with them a stick of hard wood about two feet long and sharpened at both

A MUREX ("MUREX PALMA-ROSE.") OF CEYLON.

ends. When a shark was encountered from which they could not readily
escape, they would snatch this weapon from their belts, grasp it in the
middle, and thrust it dexterously crosswise into the widely distended mouth
of the monster, opened to seize them. To shut down his jaws upon such a
skewer would undoubtedly discomfit a shark or anything else; but when
one thinks of the time, nerve, and sure aim it would require to accomplish
this feat, he begins to doubt whether it really ever was tried. I advise
you, therefore, to prove the story better than I have been able to do, before you pin *all* your faith to it.

An Australian pearl-diver, writing about this matter in "The Century"
magazine a few years ago, assures us that a fifteen-foot shark, magnified
by the water, and making a bee-line for one, is sufficient to make the stoutest
heart quake, in spite of the assertion that sharks have never been known to
attack a man in a rubber diving-dress. He adds:

Neither is the sight of a large turtle comforting when one does not know exactly what it is,
and the coiling of a sea-snake around one's legs, although it has only one's hands to bite at, is, to
say the least, unpleasant. A little fish called the stone-fish is one of the enemies of the diver.
It seems to make its habitation under the pearl-shell, as it is only when picking up a shell that
any one has been known to be bitten. I remember well the first time I was bitten by this spite-
ful member of the finny tribe. I dropped my bag of shells, and hastened to the surface; but in
this short space of time my hand and arm had so swollen that it was with difficulty I could get
the dress off, and then was unable to work for three days, suffering intense pain the while. Afterward I learned that staying down a couple of hours after a bite will stop any further discomfort,
the pressure of water causing much bleeding at the bitten part, and thus expelling the poison.

All the oysters when brought ashore are opened in vats of water, and
carefully examined for the pearls they may contain half embedded in their

mantles; but very few reward the diver with gems worth selling separately or otherwise than by weight as "seed" pearls. Many divers, therefore, do not themselves take the trouble of opening what they catch, but sell them unopened at a few cents a dozen, preferring the small and steady assured income to the chances of failure or a fortune.

The round, flat, beautiful shells are saved, and their sale (for mother-of-pearl work) brings nearly as much money into the pearl-fishing communities in the course of a season as is derived from the pearls themselves.

What beauty, as well as usefulness, have shells! And how wide is the science (conchology) that deals with them, and tells us not only their structure and manner of life, but interprets the part which their extraordinary forms, ornaments, colors, and appendages play in their "struggle for existence" down in that populous green under-world of the waters!

I know a picturesque old house [writes a charming pen in one of the early volumes of "Scribner's Monthly"] that has a many-shelved pantry devoted to the exhibition and sale of

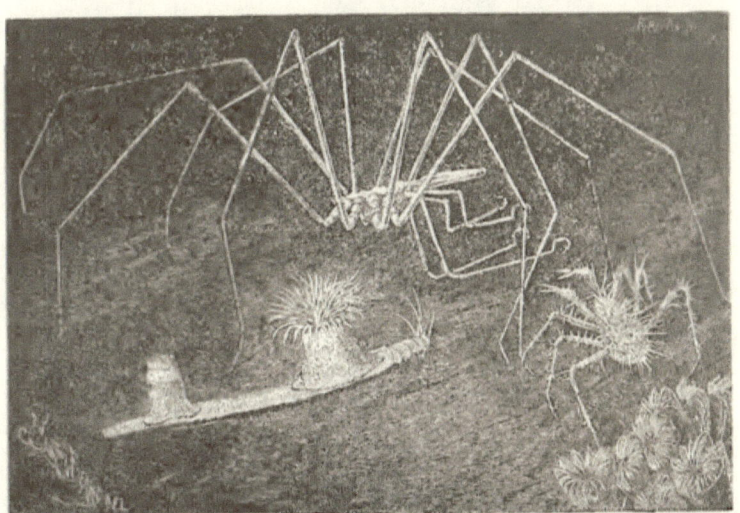

ON THE GULF STREAM SLOPE, FROM ONE TO TWO MILES BELOW THE SURFACE.

shells, collected in many a long voyage to the remotest parts of the five oceans. Apart from their scientific interest, their associations with alien races and far-off countries, how beautiful these shells are in themselves! and how readily might the prevailing vulgarities and absurdities in the decoration of glass and porcelain be corrected by studying the ceramics of nature! How, for instance, is our sense of cleanliness served and our appetite wooed by the extreme smoothness, hardness of surface, and pearly white of the oyster-shell! What decoration in the part that receives the viand, what metallizing the surface or changing it into artificial marble, or covering it up with pictures, would take the place of the pure, colorless shell?

Every species of these shells has a principle of growth, or law of form, peculiar to itself and yet based upon some more general law of form common to other species. . . . In the comb of Venus, for instance, the initial impulse of structure tends to produce a series of spines of a peculiar curvature, and arranged after a certain order that involves the use of similar curves. It is interesting to study the development of this simple principle into the complex and singular form of beauty comprised in the shell itself, the idea being carried into the most minute particulars — even the dark markings at the mouth being shaped like spines, and every small projection on the surface evidently being an arrested development of spines. In the *Murex haustellum*, on the contrary, nodules take the place of spines. In the *M. endivia* an entirely different idea is developed. Notice the cross-striations. Instead of prolonging themselves into cylindrically pointed spines, as in the case of the Venus' comb, or bunching themselves into knobs, as in the *M. haustellum*, they expand into wonderful foliated projections, the edges of which are beautifully fluted, like the leaves of the lettuce. Another fine effect is afforded by the different texture of the inside and outside surfaces, down to the smallest foliation, the inner parts exhibiting a polished pearly white, and the outside a gray and wrinkled skin. Observe that, however rough or dull of hue the outside of a shell, its lips are always pure and often flushed with lovely color; for, as a rule (and here is another hint to decorators), Nature distinguishes by some adornment the most significant parts of her creatures, where life and use are centered. . . . The ocean, indeed, beautifies all it touches. Give it any rough shard, and it will so roll it about, and lick it with its waves, and smooth it with their soft attrition, that it will return you a polished and shapely nodule, exhibiting all the beauty of color and surface of which the material is capable.

INDEX OF ILLUSTRATIONS

GENERAL INDEX